A PRACTICAL COURSE IN FUNCTIONAL PROGRAMMING USING ML

A PRACTICAL
Course in Functional Programming Using ML

RICHARD BOSWORTH

McGRAW-HILL BOOK COMPANY

London · New York · St Louis · San Francisco · Auckland
Bogotá · Caracas · Hamburg · Lisbon · Madrid · Mexico
Milan · Montreal · New Delhi · Panama · Paris · San Juan
Sao Pãulo · Singapore · Sydney · Tokyo · Toronto

Published by
McGraw-Hill Book Company Europe
SHOPPENHANGERS ROAD, MAIDENHEAD, BERKSHIRE, SL6 2QL
Telephone 01628 23432
Fax 01628 35895

British Library Cataloguing in Publication Data
Bosworth, Richard
Practical Course in Functional
Programming Using Standard ML
I. Title
005.133

ISBN 0-07-707625-7

Library of Congress Cataloging-in-Publication Data
Bosworth, Richard,
A practical course in functional programming using ML/Richard Bosworth.
p. cm.
Includes bibliographical references and index.
ISBN 0-07-707625-7
1. Functional programming (Computer science) 2. ML (Computer program language) I. Title.
QA76.62.B67
005. 13'3--dc20 95-51
CIP

1234BL8765

Typeset by MFK Typesetting, Hitchin, Herts
and printed and bound in Great Britain by
Biddles Ltd, Guildford and King's Lynn

Printed on permanent paper in compliance with ISO Standard 9706.

To Ruth

CONTENTS

PREFACE

This book is based on a one-semester course in functional programming which I have been teaching for the last five years. The course is for first-year university students with no advanced mathematical training. As a result, the pace is fairly easy for the first few chapters, speeding up rapidly as the student gains confidence. The book would be suitable for anyone who wants a leisurely but rigorous introduction to functional programming.

The course was designed to be practically based, leading to the implementation of a non-trivial prototype system, and this pattern is visible in the book, which has a wealth of practical exercises, culminating in the implementation of a library system. I have tried to apply ML to the kind of 'real-life' problems that are routinely dealt with in books on imperative programming, but which seem strangely absent from those on functional programming. As a result, some of the solutions are not particularly mathematically elegant. But I believe it is important to meet imperative programming head-on in this way, and show that a functional language can do as well or better. As a consequence, I (and my students) have had to think seriously about issues such as interactive human–machine communication, validation of input, and permanent storage of data. The programs presented here, while no doubt capable of improvement, at least demonstrate that these problems can be solved within the functional paradigm.

The key features of the book are as follows:

- A truly functional subset of ML is used throughout, without any imperative features being mentioned. This means, for example, that a pseudo-functional input/output system is used, on which standard inductive proofs can be carried out.

- Linear structures such as stacks, queues and lists are covered in some detail, but space considerations mean that more advanced dynamic types such as trees are omitted. This is a pity, as functional algorithms on dynamic types are very simple and intuitive, but in a one-semester course, not everything can be covered. The interested reader is referred to the many excellent books dealing with this topic.
- Higher-order functions are taken very seriously and are introduced as early as possible (I believe they are one of the keys to more accurate and efficient programming).
- The importance of proof is emphasized throughout, and the habit of informally proving everything that is written is inculcated from the beginning. Functional programming lends itself to this approach.
- The approach is, as far as possible, a problem-solving one — a problem is stated, then the features of the language needed to solve that problem are introduced in a natural way.
- There are many worked examples, including several non-trivial systems. In programming, as in other crafts, the student learns initially by imitation. Only gradually does he or she 'take flight' and produce original work. Eventually, the student may become more proficient than the tutor. If you think you have discovered a better solution to a problem in this book, I would be very pleased to hear from you (also if you discover an error, of which undoubtedly many remain).

Thanks are due to many people, including: Gordon Bull for initiating the first-year course, and thus introducing me to functional programming; Richard Mitchell for persuading me to write this book; Dan Simpson for giving me the time to write it, and for scrutinizing the result so carefully; my wife for her active encouragement and patience (you can have the spare room back now, Ruth); my office partners, Franco Civello and Stuart Kent, for putting up with dark mutterings in the corner; the ML development team for producing such an excellent language in Standard ML; the Poplog development team, especially Robert Duncan and Simon Nicholls, for a first-class and efficient implementation of SML; Apple Computers and Microsoft for providing comfortable tools to write this book; Bruce MacLennan for explicating the method of differences; Simon Thomson for unwittingly providing the ideas on which Chapter 10 is based; Joyce Barlow for her keen interest in all aspects of teaching and learning; the anonymous reviewers at McGraw-Hill for pointing out deficiencies, inaccuracies and plain wrong-headedness in earlier drafts; Rupert Knight for being a helpful, tolerant and understanding editor; David Turner for showing that proof of functional programs can be delightfully simple; Bill Crawford for having brilliant ideas, especially recursive selection; Nigel Brown for masterminding the conversion of UML to BUML; Ian Oliver for testing functions on a variety of ML interpreters; Stuart Corner for authoring the help files on the Brighton University ML system; and, of course, all the people to whom I've taught ML over the last five years, especially the ones who didn't understand a word I was saying, and so forced me to clarify my thoughts.

Brighton, 1995.

CHAPTER
ONE

INTRODUCTION

1.1 A PROBLEM SOLVED

This is a book about programming electronic digital computers. Almost as soon as they were invented, it was realized that the process of instructing computers to operate correctly was a task of a completely new kind, for which the history of technology provided no precedents. John von Neumann, one of the pioneers of computer design, and a great mathematician, put it well:

> The actual code for a problem is that sequence of coded symbols (expressing a sequence of words, or rather of half words and words) that has to be placed in the Selectron memory in order to cause the machine to perform the desired and planned sequence of operations, which amounts to solving the problem in question. Or to be more precise: This sequence of codes will impose the desired sequence of actions on *C* by the following mechanism: *C* scans the sequence of codes, and effects the instructions, which they contain, one by one.
>
> If this were just a linear scanning of the coded sequence, the latter remaining throughout the procedure unchanged in form, then matters would be quite simple. Coding a problem for the machine would merely be what its name indicates: Translating a meaningful text (the instructions that govern solving the problem under consideration) from one language (the language of mathematics, in which the planner will have conceived the problem, or rather the numerical procedure by which he has decided to solve the problem) into another language (that one of our code).
>
> This, however, is not the case. We are convinced, both on general grounds and from our actual experience with the coding of specific numerical problems, that the main difficulty lies just at this point.[1]

The good news that this book brings is this: the problem that John von Neumann described has been solved. The mismatch between the human view of a problem and the computer's view has disappeared; it is now possible for a computer to be instructed in terms essentially identical

Der Mann ist in dem Haus

Fig. 1.1 Two representations of the same situation

to those in which we conceive the solution to a problem. Furthermore, the kinds of computer applications amenable to this approach comprise far more than von Neumann's 'numerical problems'; most modern applications of computers can be designed in the terms we use in this book — in terms of functional programming.

If functional programming is such a giant step forward, then why isn't everyone using it? To answer this question we can look back with profit to previous eras. It is noticeable that changes in notation take a great deal of time to become established. Take, for instance, the move from ideograms to letters — a change that even now is not complete in some parts of the world.

If you consider the left- and right-sides of Fig. 1.1 you will see some of the issues involved in a change of notation. The left-hand representation is a direct, if somewhat crude, mapping of the situation as perceived by a human observer into a two-dimensional image. It can be easily grasped without special training, and might be spontaneously produced by a young child.

The right-hand representation is much more indirect and abstract. It requires familiarity with linguistic ideas such as concept ordering, mapping objects to words, using a symbol to represent containment of one object by another, grammatical rules, and so on. A great deal of training and experience is required to produce and understand the right-hand representation. Yet there can be little doubt that, for people who understand it, it is a superior representation, as anyone who has tried to decode the diagrammatic signs in an international airport will readily agree.

Or again, take the switch from Roman to Arabic numerals, a process that took some three centuries to establish itself. If we want to do a fairly simple addition operation, the Roman scheme has some advantages: for example, add twenty-three to twelve:

XXIII plus XII is XXXIIIII is XXXV

We simply take the symbols and literally put them together, applying a few simple equivalence rules, like IIIII = V, or IIII = IV, to shorten the resulting expression. Subtraction is almost as easy. The Arabic version is much more abstract:

$$\begin{array}{l}\beta\chi+\\ \underline{\alpha\beta}\\ \underline{\chi\varepsilon}\end{array}$$

(I have taken the liberty of substituting Greek letters for the usual Arabic symbols, so the reader may relish the sense of complete confusion that a medieval European would experience when confronted with this notation.) We have a host of symbols which are not connected in any obvious way to the objects they represent. We have to learn a whole set of abstract rules, such as $\alpha + \alpha = \beta$ $(1 + 1 = 2)$, plus the concept of carrying from units to tens and so on. Only after we have all this conceptual apparatus can we start to add numbers together.

Of course, once the numbers become large the Arabic scheme comes into its own, as the same rules apply and no new concepts are necessary, while the Roman scheme requires new symbols such as C and M, and new rules concerning them. When we try to multiply there is no contest: for example, multiply twenty-three by twelve:

XXIII times XII is ???

The Romans resorted to the abacus at this point, but with Arabic notation a pencil-and-paper procedure or **algorithm** is possible:

$$\begin{array}{l}\beta\chi\times\\ \underline{\alpha\beta}\\ \beta\chi\zeta+\\ \underline{\delta\phi}\\ \underline{\beta\gamma\phi}\end{array}$$

Once again we see a move away from the intuitive and direct towards the abstract and formal.

1.2 IMPERATIVE AND FUNCTIONAL PROGRAMMING

My argument is that the move from the kind of computer programming that is prevalent at the moment, so-called **imperative programming**, to the computer programming of the future, **functional programming**, is a notational change of this type. Already functional programming is the paradigm of choice for teaching computer programming in some 80 institutions around the world.[2] It is also becoming popular among computer hobbyists. It has been used on some commercial projects. The slow process of notational change is gradually taking place around us.

Given that this process is occurring, it makes sense to learn about computer programming in the most up-to-date way. This is what this book helps you to do. Once the reader has grasped the principles of programming set out in the following chapters, he or she will be in a good position to tackle other programming languages (including imperative ones). But most importantly, I believe, he or she will have a good understanding of the abstract principles involved in programming computers, and this knowledge and skill will remain relevant whatever particular language or system is used.

Let us compare a program written in a conventional imperative language and a functional language. (Of course, I have chosen an example which favours the functional approach: the calculation of a factorial number. But the points I am about to make hold for many other examples.) The factorial of a positive integer number (often written as $n!$) has a particularly simple mathematical characterization:

$$
\begin{aligned}
0! &= 1 \\
n! &= n \times (n\text{-}1)!
\end{aligned}
$$

The factorial of zero is one, and the factorial of any positive integer n is n times the factorial of the previous integer. These two equations are sufficient to specify the factorial of any positive integer number. Here is an imperative version of a program to calculate the factorial of a number:

```
function factorial (n : integer) : integer;
var f, i : integer;
begin
  f := 1;
  for i := 1 to n do f  := f * i;
  factorial := f;
end;
```

You can see that this program is a recipe for calculating a factorial, rather as one would perform it on a pocket calculator. The initial answer (1, for factorial 0) is set up in a register of the machine called f. A counter called i (also set up in a register) is used to control the course of the calculation. i is set initially to the value 1, and is then stepped through the values 2,3, . . . up to n. At each step of the calculation, the answer in f is multiplied by the current value of i, so we have

Number of steps	f
0	1
1	1
2	2
3	6
4	24
5	120
...	...

It is easy to check that for n steps, the result will be n! What is not so easy is to invent the program in the first place. As John von Neumann remarked, the mathematical formulation of the problem has to be modified radically to fit the engineering constraints of the computer. Now let's look at a functional version of this program:

```
fun factorial 0 = 1
|   factorial n = n * factorial (n-1)
;

val factorial = fn : int -> int
```

The mathematical formula has been transcribed, with a few systematic changes (* for × and so on), to give the computer code — a rather more straightforward process. Furthermore, on reading this code, the computer system immediately calculates that the function takes an integer as argument and returns an integer as result. It displays this fact to the programmer (in the underlined message underneath the function). Not only is the invention of the program an order of magnitude easier, but the simplicity and regularity of the notation allows the computer to contribute interactively to the process of design.

Of course, in the change from a more concrete to a more abstract notation, something is lost. In the case of the change from ideograms to letters, it is the direct, intuitive representation of concepts such as 'man' and 'house'. In the case of the change from Roman to Arabic numerals, it is the direct correspondence between the objects of calculation and the numerals used for the calculation. What have we lost in the change from imperative to functional programming?

One thing we lose is a model of how the calculation is carried out. For the imperative implementation of the factorial function, we can imagine the calculation clanking away on an electronic calculator, replete with registers and arithmetic circuitry. This model isn't available to us for the functional formulation. The calculation of a particular factorial takes the form of a **reduction** process in which equals are substituted for equals. So we have, for example:

```
  factorial (4 )
= 4 * factorial (3)
= 4 * (3 * factorial (2))
= 4 * (3 * (2 * factorial (1)))
= 4 * (3 * (2 * (1 * factorial (0))))
= 4 * (3 * (2 * (1 * 1)))
= 4 * (3 * (2 * 1))
= 4 * (3 * 2)
= 4 * 6
= 24
```

(The computer does some calculation as well as substitution.) This process does not translate well onto the architecture of current computers, with their fixed registers and pigeon-hole-like storage. So we also lose some efficiency (in both time taken and space used) when we adopt the functional paradigm on current computers. This effect can be over-exaggerated, however. The average computer user has, in the last few years, seen the amount of computer power and storage capacity available to him or her personally increase by a factor of a thousand or so. This is a qualitative change from a situation of famine to one of plenty. We simply don't have the constraints of processing power and storage today which plagued us up to the last decade. So we can afford to trade off the inefficiencies of functional languages (on current equipment) against the increased speed and security with which we can design functional programs.

1.3 THE DESIGN-TIME/RUN-TIME TRADE-OFF

What exactly is the trade-off in quantitative terms? It is very difficult to be precise in this area, but an experiment[3] in which an identical system was coded in a variety of functional and imperative languages indicates that, for a medium-sized application, a functional implementation will be somewhere between one-third and two-thirds as large (in terms of lines of program text) as an imperative application. Even more strikingly, development time for designing, writing and testing the application varied from one fifth to one quarter of the time required for an imperative language, depending on the functional language used. The resulting functional systems were slower than the imperative one, the difference depending on the language-implementation technique. For a reasonably efficient functional implementation, the slow-down was around two to five times.

These figures have to be taken with a pinch of salt, of course — many other factors affect the result, including programming techniques and programmer preferences — nevertheless, most people who work with functional languages would agree that large savings in program size and system development time can be made by using a functional language. These benefits appear to become more pronounced as the size and complexity of the system increases.

1.4 WHY ARE FUNCTIONAL LANGUAGES SO MUCH BETTER?

These figures (especially those for development time) cannot be totally explained on the basis of the compactness of functional code — there must be some other factors influencing the result. I believe that psychological considerations come into play here.

Computers are programmed by human beings, and computer systems serve human needs, so the kinds of systems that are produced have to make sense in human terms. One thing that human beings are very good at is abstraction and classification. We can distinguish a tree from a person, and even distinguish different kinds of trees and people. This ability does not depend on any particular sense — a blind person can tell a great deal about the kind of person he or she is talking to by the sound of their voice alone.

As well as categorizing objects, we also categorize processes, such as walking, singing, writing and so on. We also qualify these categories: walking the dog, singing an oratorio, writing a book. Suppose we decide to walk the dog one evening. We, could express this in abstract terms as

$$walk(dog)$$

In functional programming terms, we say the **function** *walk* takes the **argument** *dog*. But more than this, the process of walking the dog changes things, in particular the dog — he now no longer needs a walk. The process of walking has transformed the dog. We could express this as

$$tired\_dog \quad = \quad walk(dog)$$

The function *walk*, taking the argument *dog*, also produces a **result** *tired_dog*, which is equivalent in some way to the expression *walk*(*dog*). In this simple notation, we have captured the essence of (an aspect of) walking a dog. We have described a **mapping** (Fig. 1.2).

The view taken by current theories (for example, that of Gerald Edelman[4]) is that the human mind conceptualizes situations like this in terms of mappings between groups of neurons. (By contrast, an imperative view of the situation would be that a portion of our brain devoted to

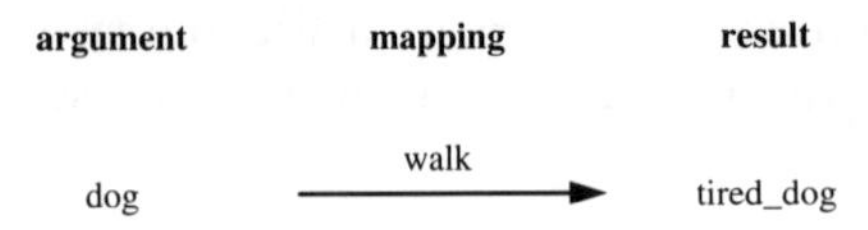

Fig. 1.2 Walking the dog

the concept 'dog' has its contents altered, by some agency, from 'lively' to 'tired' — just like a register in a calculator.) Whatever the truth of the matter, it is clear that a mapping represents a good abstract model for a process, and one that has a natural appeal for human beings.

Another feature of the human mind is its inability to deal with too many things at once. We are *attentive* creatures, always attending to something, perhaps not that to which we should be attending. William James, the nineteenth century psychologist, called this 'the taking possession by the mind, in clear and vivid form, of one out of what seem several simultaneously possible objects or trains of thought'. When we design computer programs, we like to direct our attention to each aspect of the problem in turn. In the case of computer systems, these aspects correspond to the various processes going on inside the computer, and the various data on which they act.

As a simple example, suppose we wish to write a program to take a list of numbers, and double every number in the list. We could conceptualize this as

$$double\_list(list\_of\_numbers)$$

or perhaps

$$doubled\_list\_of\_numbers \;=\; double\_list(list\_of\_numbers)$$

This is the goal we are aiming for. Now let us do a spot of design on this program. We can separate out from our overall program concept two subconcepts — the idea of doubling a number, and the idea of doing something to every number in a list of numbers. Let us direct our attention to each of these concepts in turn. First, doubling a number will be a function

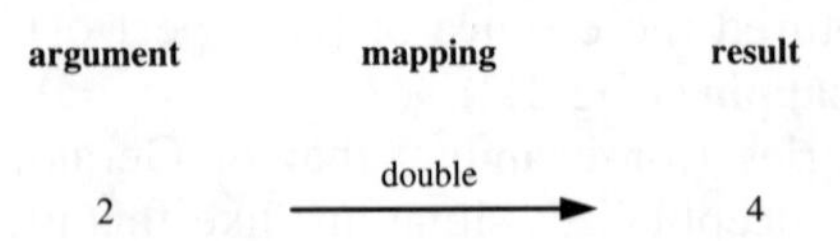

Fig. 1.3 Doubling a number

taking an integer as argument and producing an integer as result. Let's call it *double* (Fig. 1.3).

Now for the idea of doing something to every item in a list. This too can be thought of as a function — a function which takes another function such as *double* as argument and produces a new function, which operates on a list of numbers, as result. Let's call this process *listifying* (Fig. 1.4). Now we have

$$double\_list = listify(double)$$

and so the design of our program is finished — all that's left is to actually write the functions *double* and *listify*! We're not in a position to do this yet, but I hope you can see how the process of designing a functional program involves directing our attention to each aspect of the problem in turn. Of course, this has the corollary that when we come to design a similar program, we may very well find that we have already written many of the functions we need. And again, when we wish to modify some aspects of our original program, we can find the functions that represent these aspects, and modify them, without having to worry about the rest of the program. As a large program may consist of hundreds or thousands of functions, this is a useful property for a program to have.

Imperative programs are notorious for not having this property, each change bringing a harvest of new 'bugs' to be fixed:

> Big progs have lots of bugs; programmers try to smite 'em,
> But these bugs breed yet more bugs, and so on *ad infinitum*.

While on the subject of errors, let's admit that human beings have a natural desire to do things correctly, and avoid making mistakes. We shall see that as part of the process of writing a function, an informal proof is carried out that the function is correct. In this book every function that is written is informally proved correct before a word of code hits the page. The proof is embedded in the code and can be rehearsed at any time for the programmer's own satisfaction or that of others. For imperative programs, alas, proofs are quite separate from the written code and are

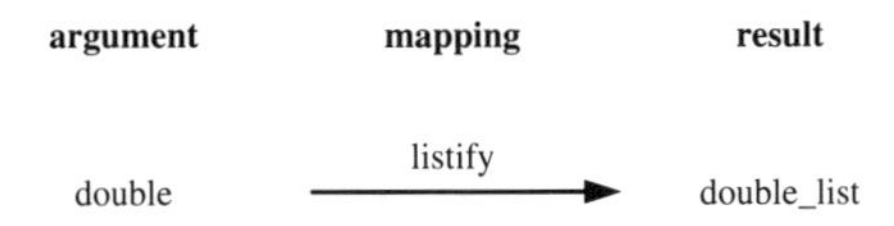

Fig. 1.4 Listifying a function

usually performed after the event, if at all, the design process resembling the behaviour of Spike Jones in 'The Man who Found the Lost Chord'.*

This emphasis on proof should not be taken to mean that functional programs are developed in some airy-fairy academic world which has no contact with reality. On the contrary, the interactive, piecemeal development of functional programs allows each function to be tested on real-world data as it is written. This intimate intertwining of proof (the function is correct) with testing (the function is appropriate) leads to programs in which we can have a high degree of confidence.

1.5 A BRIEF HISTORY

Functional notation is a relatively modern invention. It was Leonhard Euler[5] in the eighteenth century who first wrote *sin.x* for the sine of *x*. Later the dot was dropped to give *sin x* or *sin*(*x*). The link with computing goes back a long way — in the early years of this century, long before electronic computers were a reality, Whitehead and Russell[6], while attempting to provide a logical basis for mathematics, invented a notation which evolved over the years to the **lambda notation** of Alonzo Church[7]. Here is an example of lambda notation:

$$\lambda x.x+x$$

This represents a function that takes a number and returns that number plus itself — in other words, our function *double* from the previous section. The lambda calculus, based on this notation, was used by Church and others in obtaining a number of results in mathematical logic in the 1930s and 1940s — one of the most famous being the proof of the equivalence of the (functional) lambda calculus and Alan Turing's notion of an imperative **Turing machine**[8] — the basis of the claim that functional programming and imperative programming are equivalent in power.

Church's thesis that the set of computable functions is the same as the set of general recursive functions (a recursive function is one that refers to itself directly or indirectly — like *factorial* above) underlies all functional programming. In fact it is perfectly possible to learn about functional programming by starting with the lambda calculus.[9] The lambda notation itself still finds a niche in modern functional languages in the concept of an anonymous function. Here is an ML version of the above lambda expression:

```
fn (x:real) => x + x;
```

*I do not deny that many distinguished computer scientists have produced formal or semi-formal methods for designing imperative programs. What I do deny is that anybody actually uses these methods except in the most extreme circumstances.

You can see that ML is more pernickety than the lambda notation in insisting that the function only operates on real numbers!

The early decades of electronic computing were marked by a turn away from mathematical ideas, curiously enough. The first machines were so slow and restricted in storage that all efforts were directed towards using the computer's precious power as effectively as possible. This meant programming in a notation close to the internal architecture of the machine — so-called **assembly language**. There was a different assembly language for every different kind of computer (or series of computers), making transfer of programs and sharing of ideas difficult if not impossible. Worse still, each brand of computer had its associated priesthood of programmer-gurus, who were paid large sums of money to instruct the machines using these inscrutable codes and to keep them running when the inevitable bugs appeared. The chauvinism of those dark days still finds an echo in the bitterly fought 'language wars' of the present.

A ray of light appeared in 1956 with the invention of Fortran[10] (FORmula TRANslation), an attempt to move programming away from the machines and nearer to mathematical notation as used by scientists and engineers. So a user, knowing some mathematical formula such as

$$r^2 = x^2 + y^2$$

might write a Fortran 'statement' such as

```
R = SQRT( X * X + Y * Y)
```

The machine would evaluate the expression (given values for X and Y) and return a result for R. Although Fortran was a great leap forward from assembly language, and was the first *lingua franca* of computing, it institutionalized certain aspects of contemporary computer usage which, in retrospect, can be seen to have had an unfortunate influence on future developments.

The first of these was the concept of a **subroutine**. This is a pseudo-function, which has two extensions to its behaviour: it is allowed to modify the values of its arguments, and it is allowed to have implicit arguments (which may be modified by the subroutine, of course). So instead of saying

$$j = factorial(i)$$

and substituting a certain value for i to find the corresponding j, the Fortran programmer could write

```
CALL FACTORIAL(I)
```

a statement which would calculate the factorial of I, and then modify I to have the value of its factorial. Even worse, there was nothing to stop the programmer writing

```
CALL FACTORIAL
```

a subroutine 'call' which would take its value from some unspecified machine register and return a result in some unspecified register (possibly the same one). You can see that such a corruption of the idea of a function makes reasoning about programs extremely difficult. But there is a particularly dire consequence for repetitive calculations (like those needed to produce a factorial). The simple way to express this is in terms of a function which refers to itself, as in the functional program above. But a *subroutine* which refers to itself, at the same time as it modifies implicit arguments, is a mathematical nightmare. Fortran wisely forbade such self-reference by subroutines.

How, then, to cope with repetition? Fortran institutionalized the assembly language idea of a 'program loop' in the notorious DO-statement, a kind of indexed portion of program code inside which variables were repeatedly modified in a systematic way. Inside a DO-loop, you might see a fairly recognizable mathematical expression such as

```
Y = SIN(X)
```

along with such oddities as

```
J = J + 1
```

As a piece of mathematics, the latter expression is simply false: the (finite) value J can never be equated to the value J+1; J cannot be substituted for J+1 in expressions, and so on. Such a statement has a meaning in Fortran, however, it means 'increment the value of the register J by 1'. Again, reasoning about a program containing such notational quirks, though possible, is not straightforward.

These deficiencies in Fortran were recognized, but restrictions of processing power and storage capacity continued through the 1960s and 1970s and into the 1980s, only finally being alleviated by the ready availability of cheap microchip components in that decade. As a result, programming languages continued to emphasize efficiency of implementation rather than elegance or ease of understanding. These imperative features of Fortran were perpetuated in later languages, and are still causing trouble in modern languages such as Ada[11] and C++.[12]

A similarly well-meant attempt to make programming more accessible, the English-like Cobol[10] (Common Business Oriented Language), rebounded in its inventors' faces by creating a new arcane priesthood with its own hermetic rites and incantations. The failure of a 'common-sense' language like Cobol to make computing accessible to the multitude should have highlighted for ever the difference between informal 'natural languages' such as English and the kind of formal notation needed for computer programming.

At the same time that Cobol and Fortran were exerting their hegemony over the commercial and scientific computing worlds, a Stanford computer scientist called John McCarthy was inventing a functional programming language based on Church's lambda calculus. Lisp[13] (LISt Processing language) was originally designed to do symbolic differentiation on mathematical formulae, a task for which other languages of the time were woefully inadequate. It is now in use all over the world on a wide range of applications, but the Lisp in use today shares with McCarthy's original language little more than the plethora of brackets that are its trademark. Over the years, worries about efficiency have converted Lisp from a purely functional language to a mish-mash of imperative and functional features. Once again, short-term expediency has triumphed over traditional mathematical concerns such as elegance and clarity.

The third act in this tragedy is being played out in our time. Having made the terrible error of re-inventing functions as subroutines, the computing community is currently committing another monumental blunder by attempting to re-invent abstract algebra as 'object orientation'. This time we haven't even the excuse of expensive and slow hardware as justification for our spurning of mathematical tradition.

An algebraic system is a pleasantly self-contained mathematical structure consisting of a set of values and a set of operations on those values, each operation producing a value in the set as a result. For example, the integers . . .,-3,-2,-1,0,1,2,3,. . . form a set of values on which we can perform operations such as addition and multiplication. We can find general properties of algebras and prove general results about them. The concept of an algebra translates well into the computing world, where it is known as an Abstract Data Type.

Unfortunately, the computing community has decided to add an implicit 'state' to abstract data types to give 'objects'. Now the *same* operation on an object can give a *different* result, depending on its state. For example, it would be perfectly in order to have an object *Integer*, in which addition was affected by the state. An instance of *Integer* such as the number 2, could have the following property:

$$2 + 1 = 3 \text{ sometimes, but on other occasions, } 2 + 1 = 4$$

Fortunately, another strand of computing development has kept faith with the idea of building programs which make sense. Starting with (original) Lisp, and moving through ISWIM,[14] APL,[15] SASL,[16] FP,[17] original ML,[18] HOPE,[19] Miranda,[20] Standard ML[21] and Haskell,[22] this strand of development has eschewed imperative features as far as possible and utilized traditional mathematical ideas such as abstraction, generalization, explicitness and elegance.

With computer processing power and storage becoming cheaper all the time, and with novel computer architectures becoming easier to implement, the future now looks bright for functional programming.

1.6 AND FINALLY . . .

I learned programming, many years ago, using a long-dead imperative language called Algol 60. At the time this language was considered at the forefront of software technology. It certainly enabled me to program at a level of abstraction far above that which was normal at the time. It is my hope that the reader of this book will similarly learn to conceptualize programming in a way which is far in advance of current norms.

My mind is warped by decades of imperative programming — yours is free to use the new concepts of functional programming. Go to it — and may all your programs be beautiful!

1.7 CHAPTER SUMMARY

This entire chapter has been devoted to selling the concept of functional programming to the hapless reader. Claims have been made that functional programming has a longer pedigree than imperative programming; that it corresponds more precisely to the way the human mind operates; that it provides a way of drastically reducing program development times and program size; that functional programs are inherently easier to prove and test and contain fewer errors; that they are easier to modify and update without compromising their integrity. A biased version of the history of computing was also presented. The reader is recommended to study the rest of the book closely to check whether there is anything of substance in these claims.

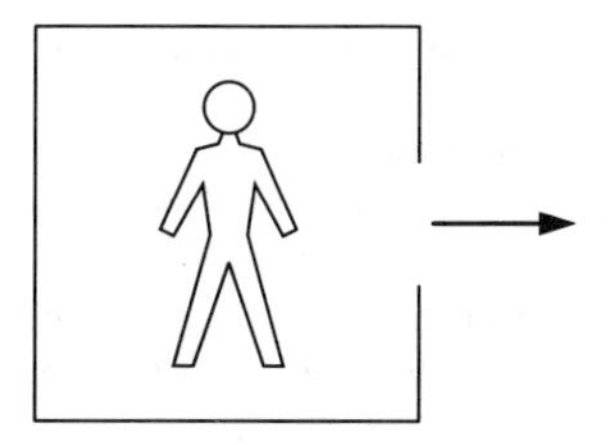

Fig. 1.5 Curious symbol observed in airports

EXERCISES

1. Try to think of as many different meanings as you can for the curious symbol shown in Fig. 1.5 which is seen in international airports.
2. Try to think of some examples of a change of notation which has produced a significant effect.

REFERENCES

1. Goldstine, H. H. and von Neumann, J. *Planning and Coding of Problems for an Electronic Computing Instrument*, Part II, Vol 1 of a Report prepared for US Army Ord. Dept., 1947.
2. Appel, A. W., Dept of Computer Science, Princeton University, 1994. Result of an informal survey on the Internet.
3. Sanders, P, 'An evaluation of functional programming for the commercial environment', in *Mathematical Structures for Software Engineering*, De Neumann, Simpson and Slater (Eds), IMA conference series, new series, 27, Oxford, 1991.
4. Edelman, G., *Bright Air, Brilliant Fire: on the matter of the mind*, Basic Books, 1992.
5. Euler, L., *Introductio in analysin infinitorum*, Lausanne, 1748.
6. Whitehead, A. N. and Russell, B., *Principia Mathematica*, Cambridge University Press, 1910.
7. Barendregt, H. P., The *Lambda Calculus: its syntax and semantics*, North-Holland, 1984.
8. Turing, A. M., 'On computable numbers, with an application to the *Entscheidungsproblem*', *Proceedings of the London Mathematical Society*, Series 2, **42**, pp 230-265, 1936.
9. Michaelson, G., *An Introduction to Functional Programming Through Lambda Calculus*, Addison-Wesley, 1988.
10. For the history of Fortran and other early languages see Sammet, J., *Programming Languages, history and fundamentals*, Prentice Hall, 1969.
11. The latest proposed standard for Ada is Ada9X. A good introduction to the language is Skansholm, J., *Ada from the Beginning*, 2nd edn, Addison-Wesley, 1994.
12. The standard work is Stroustrup, B., *The C++ Programming Language*, Addison-Wesley, 1986, but there are many other good textbooks.
13. McCarthy introduced the language in MacCarthy, J., 'Recursive functions of symbolic expressions and their computation by machine', *Communications of the ACM*, April 1960. The most-used dialect today is Common Lisp.
14. Landin, P. J., 'The next 700 programming languages', *Communications of the ACM,* **9**, 3, March, 157-166, 1966.

15. Iverson, K. E., *A Programming Language*, Wiley, 1962, is the original reference. A number of textbooks are available.
16. Turner, D. A., *The SASL Language Manual*, University of St Andrews, 1976.
17. Backus, J. W., 'Can programming be liberated from the von Neumann style? A functional style and its algebra of programs', *Communications of the ACM*, **21**, 613-41, 1978.
18. Gordon, M. J. C., Milner, R., Morris, L., Newy, M. C. and Wadsworth, C. P., 'A metalanguage for interactive proof in LCF', *Proceedings 5th ACM Symposium on Principles of Programming Languages*, Tucson, 1978.
19. Burstall, R. M., MacQueen, D. B., Sanella, D. T., *Hope: an experimental applicative language*, CSR-62-80, Department of Computer Science, University of Edinburgh, 1980.
20. Turner, D. A., 'Miranda — a non-strict functional language with polymorphic types', *Proceedings of Conference on Functional Programming Languages and Computer Architecture*, Nancy, 1985. LCNS 201, Springer-Verlag. The book by Bird and Wadler mentioned in the Bibliography uses a language very similar to Miranda, and is a good introduction.
21. Milner, R., Tofte, M., and Harper, R., *The Definition of Standard ML*, MIT Press, Cambridge, Massachusetts, 1990, is the ultimate reference, but is not recommended for beginners. See the Bibliography for a selection of books on ML.
22. Davie, A. J. T., *An Introduction to Functional Programming using Haskell*, Cambridge University Press, 1994, is a recent textbook.

CHAPTER

TWO

FUNCTIONS

2.1 INTRODUCTION

In general, 'what is a . . . ?' kind of questions are not very enlightening; we can only express the answer to 'what is an A?' in terms of other concepts (B, C, etc.). We then have the questions 'What is a B?', 'What is a C?' This phenomenon, well known to parents of small children, is called **infinite regress**. So we shall not attempt to define what a function is here. We shall show how functions are used, and two rather complementary ways of thinking about functions. The first way is pictorial (Fig. 2.1).

In this view, a function is a **mapping** from an argument type to a result type. The mapping is **many-to-one**. This means that we can express the result of applying the function to its argument using the familiar nota-

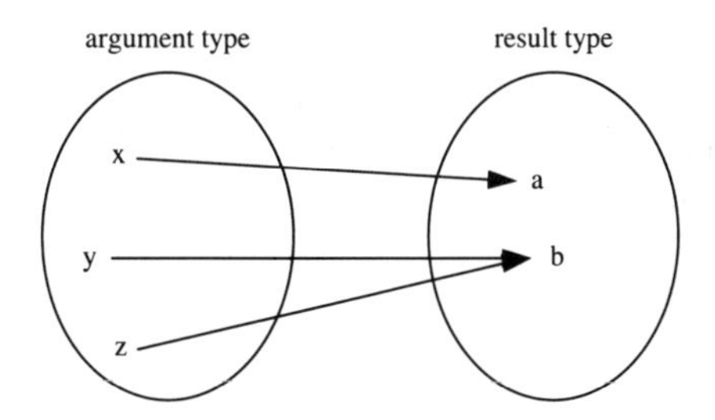

Fig. 2.1 A function

tion *f(x), f(y), f(z)* and so on, and this result is unambiguous. So for the function *f* above:

```
f (x) = a
f (y) = b
f (z) = b
```

There is no reason why we can't leave out the parentheses and say

```
f x = a
f y = b
f z = b
```

As we have defined it here, the argument type and result type are part of the function. The function *f* carries its argument and result type around with it as baggage. Is this excess baggage? Not really, as both argument type and result type are needed to delineate what a function can do. The argument type tells us exactly which values we can feed to the function (for the function *f* above, *x*, *y* and *z*), and the result type tells us exactly which values the function will return to us (*a* or *b*). Both these pieces of information are essential to us if we want to use the function for practical information processing.

We want to avoid the situation where we supply the function with a value it cannot digest, and equally we want to be sure that the values the function returns to us are within the range of what we expect, especially as we may be feeding these values in turn to another function.

2.2 TYPES

We haven't yet stated precisely what a type is. A type clearly bears some relation to a set in mathematics. We can think of the standard mathematical sets of numbers:

$\mathbf{N}_1$ the set of (Greek) **natural numbers**: 1,2,3,4 . . .
$\mathbf{N}_0$ the set of (Arabic) **natural numbers**: 0,1,2,3,4 . . .
Z the set of (German) **integers**:
. . . ,-3,-2,-1,0,1,2,3, . . .
(**Z** comes from *zahlen*, the German word for counting)
Q the set of (English) **rationals**, that is, numbers of the form a/b, where a and b are integers. (Computers do all their arithmetic using rational numbers. Programmable computers were invented by Charles Babbage in England.)
R the set of **real numbers** — so called because they are impossible to represent as a finite series of digits, in general. Contains members such as $\sqrt{2}$.
I The set of **imaginary numbers**. Contains members such as $\sqrt{-1}$.

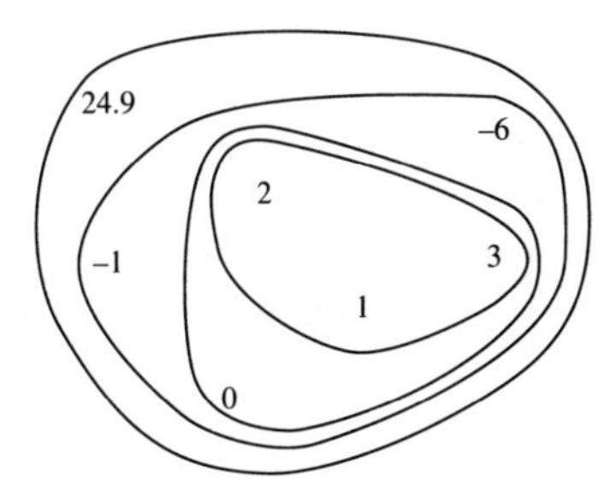

Fig. 2.2 Sets from a hierarchy

C The set of **complex numbers**. Contains all the real and imaginary numbers, and numbers which are a mixture of both.

It is clear that these sets form a hierarchy, which was built up historically as more and more numbers were admitted into the fold (Fig. 2.2). In mathematical notation:

$$\mathbf{N_1} \subset \mathbf{N_0} \subset \mathbf{Z} \subset \mathbf{Q} \subset \mathbf{R} \subset \mathbf{C}$$
$$\mathbf{I} \subset \mathbf{C}$$

The situation with types is different. In ML, we have types corresponding to some of the standard sets above, for example the type *int* corresponds to **Z**, and the type *real* corresponds to **Q**. But each particular value in our universe of values is given precisely *one* type. (We can imagine a (non-ML) function *type* which assigns this type so that *type* (2.765) = *real* and so on.) This means, for example, that the values 3 and 3.0 have different types — *integer* and *real* respectively.

The result of this restriction is that ML types are **disjoint**; they have no values in common, so there is little point in trying to form the intersection of two types as we can with two sets (Fig. 2.3). The advantage of the

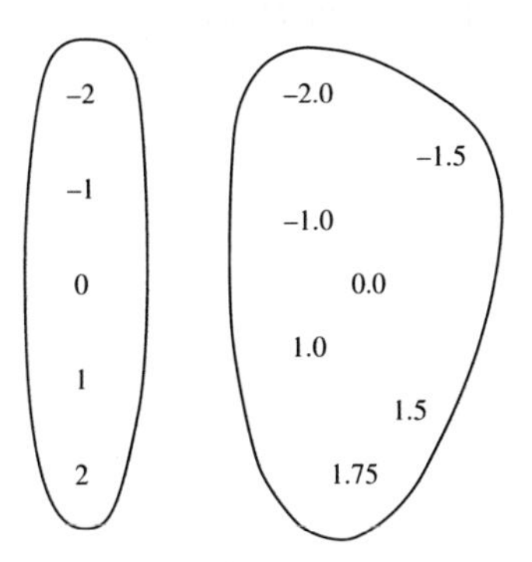

Fig. 2.3 Types are disjoint

arrangement is that values are essentially self-typing; given a value, we can uniquely determine its type.

In ML, every valid expression that you can write is **well typed**; that is, it has a well-defined type. This type may be very complicated, but it exists. This property of ML (and other **strongly typed** computer languages) means that a whole class of errors, caused by trying to apply a function to a value it cannot deal with, are excluded.

Example: Function *double*

As an example of ML's attitude towards types, let us try entering a few simple expressions into the ML system and look at the results:

```
2;

val it = 2 : int
```

The first line is our input to ML, the second line is the result. Our input is terminated by a semicolon to tell the ML system that we want the expression to be evaluated. ML recognizes the expression as a simple integer number, and tells us that 'it' (which is the name for any otherwise anonymous expression in ML) has the value 2, which is of type *int*.

This style of working is called the **read-evaluate-print paradigm**. This is just a fancy name for the idea of typing an expression into the computer for the ML system to read, instructing the system to evaluate it, and letting it print out the result. The result isn't actually underlined, of course (at least, not in my system); this is just a typographical convention to help you to distinguish what the user is saying from what the computer is saying. The ML system may output a 'prompt character' to tell you when it is ready for you to type in an expression; I have omitted this character in the conversations with ML in this book.

Now let's try a more complicated expression:

```
2 + 3;

val it = 5 : int
```

ML evaluates the expression and its type. You can see that ML has no problems with the idea of a function (+) which takes more than one argument, and which appears *between* those arguments in an expression. Let's try one more:

```
4.0 + 5;

ML TYPE ERROR - Type unification failure
WANTED    :  real * real
FOUND     :  real * int
```

This time we have not been so lucky. Although the expression makes perfect sense to us, as far as ML is concerned it is malformed, because the two arguments to the + function should have the same type. In other words, in ML only integers can be added to integers, or reals added to reals. We try again:

```
4.0 + 5.0;

val it = 9.0 : real
```

Success this time. Now let us invent a function that doubles its argument. The multiplying function $\times$ is represented by * in ML:

```
fun double n = 2 * n;

val double = fn : int -> int
```

ML has already worked out the type of the argument and result of this function! How has it done this? Simply by noting that the constant 2 has type *int*; from this it follows that x must also be an integer (since we are multiplying integers), and the result must be an integer too. So the function *double* must take an integer as argument and return an integer as result. ML expresses this using the notation *int -> int*.

From these examples we can see that ML's type system is stricter than ordinary mathematics. The disadvantage is that you have to be careful when you write out mathematical expressions. The advantage is that every value you write has a well-defined type that both you and ML can derive from the expression itself.

Now let's try applying our newly invented *double* function:

```
double 4;

val it = 8 : int

double 4.0;

ML TYPE ERROR - Type unification failure
WANTED    :  int
FOUND     :  real
```

ML will only allow the function to be applied to an argument of the correct type.

Without realizing it, we have demonstrated the second way of thinking about a function; as an **algorithm** which computes a result from a given argument. When we say

```
fun double n = 2 * n;
```

we are defining the function *double* using a recipe or algorithm which is written in terms of already known functions — in this case the multiplication function which is one of the predefined functions of ML.

You can see that this way is complementary to the first way; it would be impossible to list all the arguments and results of function *double* because there is an infinite number of each. On the other hand, it is possible to imagine a function which maps arguments to results in a very irregular fashion (for example, a function that given a person's payroll number, returns that person's salary), where an algorithmic description would be impossibly convoluted, and therefore inappropriate. We shall learn how to invent both kinds of functions.

Before leaving the subject of types, let's look at some of the limitations of ML's typing scheme. Suppose we want to write a function *square* which returns the square of its argument. By analogy with *double*, it is easy to invent the function:

```
fun square n = n * n;
```

When we enter our definition into the ML system, we obtain the reply

```
ML TYPE ERROR - Cannot determine a type for
overloaded identifier
INVOLVING:   * : 'ty1 * 'ty1 -> 'ty1
```

Because the multiply function in ML is defined to operate *either* on two integers or two reals, ML cannot decide which is meant in this case. We can overcome this shortcoming by explicitly stating the type of the argument:

```
fun int_square (n:int) = n * n;

val int_square = fn : int -> int


fun real_square (x:real) = x * x;

val real_square = fn : real -> real
```

Fortunately, cases like this where an explicit **type constraint** is necessary are few and far between.

2.3 PARTIAL FUNCTIONS

Up to now we have tacitly assumed that we can always define the function that we need, either by listing the possible values of its arguments and their corresponding results or by inventing an algorithm that is guaranteed to give the correct result. For example, suppose we want a function that negates any (integer) argument given to it. We can easily invent this function in ML:

```
fun negate n : int  = ~n;

val negate = fn : int -> int
```

(Note that minus applied to one number, so-called **unary minus**, is ~ in ML.) We now subject it to tests:

```
negate 3;

val it = ~3 : int


negate ~22;

val it = 22 : int


negate 0;

val it = 0 : int
```

But now suppose we want to find the square root of a number. There is a standard predefined ML function to do this called *sqrt*. We can find its type by giving its name to ML:

```
sqrt;

val it = fn : real -> real
```

Let's try some tests on it.

```
sqrt 4.0;

val it = 2.0 : real


sqrt 2.0;

val it = 1.414214 : real


sqrt ~4.0;

ML EXCEPTION: Sqrt
```

ML cannot give an answer because the square root of a negative number is not defined, so an **exception** message appears instead. There are two ways of looking at this situation: we could say that the argument type of *sqrt* is the positive real numbers only, but this is not a standard type in ML, so we have to discard this option. The other option is to say that *sqrt* is a function on the real numbers which is only defined for a subset of its argument type; it is a **partial function**.

The function in Fig. 2.4 is **partial**, i.e. it is not defined for all values of its argument type. The subset of the argument type that the function *can* deal with is called its **domain**. A function that is defined for every possible value of its argument type is called a **total** function.

Partial functions, in computational terms, are algorithms that cannot deal with all their possible inputs. In general, this is an undesirable situation (imagine a controller for a nuclear plant which collapsed when fed unusual values for temperature and pressure), so we try always to produce functions which are total if we possibly can.

ML provides an **exception** mechanism to ensure that applying a partial function to its argument is always a well-defined operation, even if it

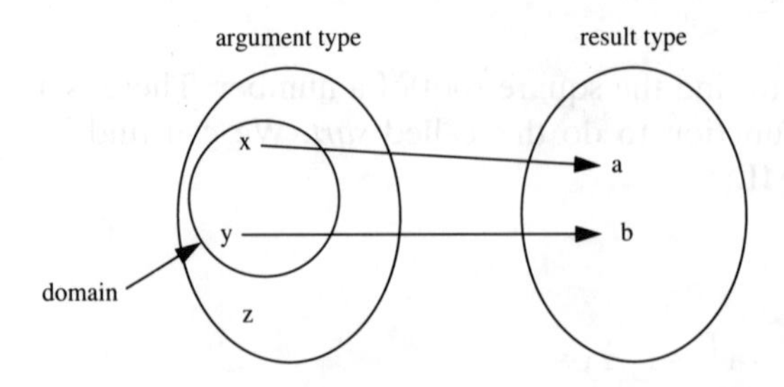

Fig. 2.4 A partial function

doesn't produce a result. We shall learn later how to raise an exception, and how to deal with exceptions that are raised by functions.

2.4 RANGE OF A FUNCTION

A function may have a limited **range** (Fig. 2.5), i.e. it may not be able to produce all possible values of its result type. This is nothing to worry about, because we can still use the result of function *f* as argument to another (total) function *g*, with complete confidence, as long as *f*'s result type is the same as *g*'s argument type. So, unlike a restricted domain, which is a dangerous situation, a restricted range is no problem at all. There is no special name for a function with a restricted range, but a function whose range is equal to its result type is called a **surjection** or sometimes an **onto function**. In some books, a function's result type is called its **codomain**.

2.5 TYPE OF A FUNCTION

By convention, the type of an expression is indicated by suffixing the expression with a colon and a type-name, e.g.:

```
4 : int                    7.6 : real

"fred" : string            [1,2,3]  :  int list
```

This convention can be extended to functions, so that a function *f* which takes a real argument and produces an integer result would have the type

```
f  : real -> int
```

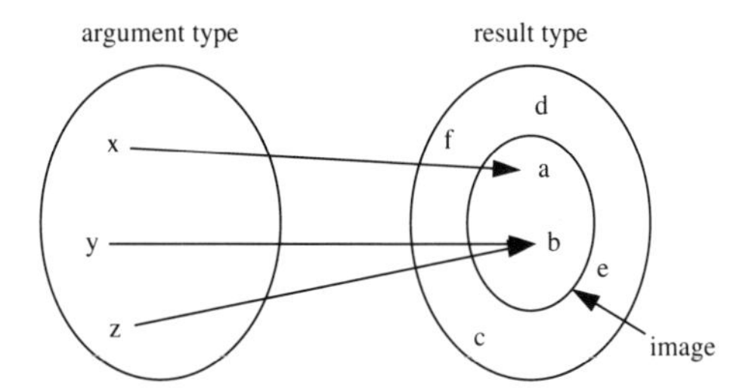

Fig. 2.5 Range of a function

We have seen this convention in use in the examples above, and we will use it freely from now on. The symbol $\rightarrow$, which we can pronounce 'arrow' or 'to' is formed in ML from the two keyboard characters *minus* (-) and *greater than* (>).

2.6 KINDS OF FUNCTION

Figure 2.6 represents a function which is **injective**, that is, each value of the argument type gives a **different** value of the result type. Another name for this is a **one-to-one** function. If a function f is injective, then it has an **inverse** function f^{-1} where the direction of the arrows is reversed (Fig. 2.7). If it is also surjective, then this inverse function will be total.

If	`f`	:	$\alpha \rightarrow \beta$
then	f^{-1}	:	$\beta \rightarrow \alpha$

Note that a function which is injective and surjective is known as a **bijection** or **bijective function.**

A special kind of function is the **identity function**, usually given the name I. This is a function that has the magical property of leaving its argument exactly as it found it, so $I(32) = 32$, for instance. Even this apparently useless function has its part to play in our study of functional programming.

Question What is the type of the identity function?

Answer The type of its result will be the same as the type of its argument. We can express this by saying it has type $\alpha \rightarrow \alpha$.

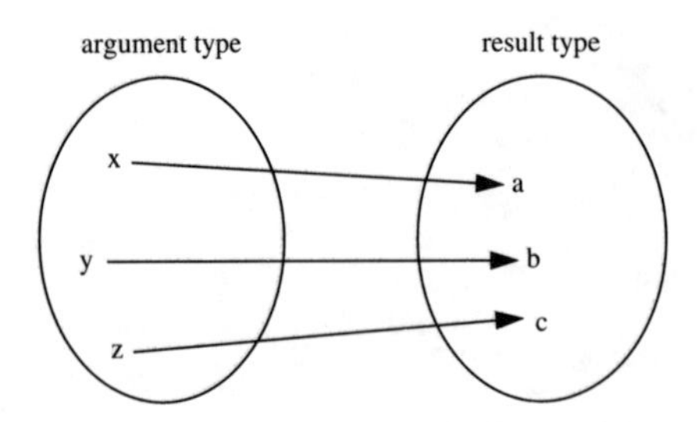

Fig. 2.6 Injective function

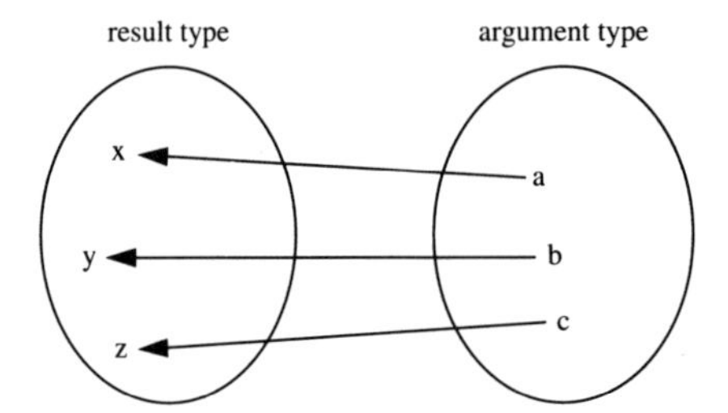

Fig. 2.7 The inverse of the function in Fig. 2.6

2.7 SYNTAX INTRODUCED IN THIS CHAPTER

This section, which you will find at the end of most chapters, is designed to help you check the format of your ML programs. **Syntax** is just the official word for the 'grammatical rules' of a language — the allowable ways in which you can put components together to make larger components, and, eventually, programs.

The syntax of ML is described using a modified version of Backus-Naur Form, or BNF to the cognoscenti. This is a metalanguage used to describe other languages. To avoid confusion about when we're talking BNF and when ML, BNF has some **metasymbols** which do not appear in ML itself. They are:

::= meaning 'is defined as'

⟨ ⟩ which are brackets enclosing optional phrases.

A typical BNF definition would be

fundec ::= `fun` *var atpat* ⟨:*ty*⟩ = *exp*

which gives the format for a function declaration. Alternative definitions are listed underneath each other. To avoid the infinite regress mentioned at the beginning of this chapter, all definitions eventually reduce to so-called **terminal symbols**, like `fun` or `=`, which are actual symbols of the language under discussion.

Let us check an ML function declaration against the above definition. Here is a function we declared earlier:

```
fun negate n : int  = ~n;
```

By comparing with the BNF definition, we can see that this is syntactically correct if `negate` is a valid *var*

`n` is a valid *atpat*

`int` is a valid *ty*

`~n` is a valid *exp*

Looking at the definitions below, *var* is defined as *id*, and one definition of *id* is 'any sequence of letters or digits starting with a letter', so `negate` is OK. `n` is a *var* too, which is one definition of *atpat*. `int` is a possible *ty*. The checking of `~n` takes a little longer, going through the stages *infexp* and *appexp* to postulate that

`~` is a valid *appexp*

`n` is a valid *atexp*

As both *appexp* and *atexp* eventually reduce to *id*, and both `~` and `n` are valid *id*s, `~n` is a valid *exp*.

In this way we have shown that our function definition is syntactically correct. However we have not shown that it is meaningful (or **semantically correct** in the jargon). We have not even shown that the types of the various components are correct. The definitions

```
fun negate n : int = ~m;
```

and

```
fun negate n : int = ~ "wrong"
```

are also syntactically correct! But no syntactically incorrect program will be accepted by the ML system, so syntactic correctness is a useful first check on our program. Here then are the syntax definitions for the Ml constructs we have used so far.

Expressions

exp	::=	*infexp*	
		exp : *ty*	typed expression
ty	::=	`int`	
		`real`	
		`string`	
		ty -> *ty*'	function type expression

infexp	::=	*appexp*	
		$infexp_1$ *id* $infexp_2$	infix expression
appexp	::=	*atexp*	atomic
		appexp atexp	application expression
atexp	::=	*scon*	special constant
		var	value variable
scon	::=	*intcon*	integer constant
		realcon	real constant
		stringcon	string constant
intcon	::=	any non-empty sequence of digits, possibly preceded by a negation symbol (~)	
realcon	::=	an integer constant followed by a point (.) and one or more digits	
strcon	::=	a sequence, between quotes ("), of zero or more printable characters	
var	::=	*id*	identifier
id	::=	any sequence of letters or digits starting with a letter	
		+	
		-	
		*	
		~	

Patterns

atpat	::=	*scon*	special constant
		var	variable

Function declaration

fundec ::= `fun` *var atpat* ⟨:*ty*⟩ = *exp*

2.8 CHAPTER SUMMARY

A function can be thought of in two ways: as a static mapping from an argument type to a result type, or as a dynamic algorithm operating on values of the argument type and computing values of the result type. In both cases, there is a unique result for each value of the argument type. Types are collections of values like sets, but types in ML are disjoint, so that each value has precisely one type.

ML's strong typing means that certain classes of error are eliminated, and types of complex expressions can be unambiguously inferred, but the price paid for this is a certain fastidiousness about arithmetic expressions.

Some functions cannot cope with all the values in their argument type; they are partial functions. ML provides an exception mechanism to deal with this situation.

Functions have types just like other values. Function types look like $\alpha \to \beta$, where α is the argument type, β is the result type, and the $\to$ symbol is pronounced 'arrow' or 'to'.

If a function can give any value in its result type as a result, it is called a surjection. If results for different arguments are different, it is called an injection. If it has both these properties, it is a bijection. A bijective function of type $\alpha \to \beta$ has an inverse of type $\beta \to \alpha$.

EXERCISES

1. Say whether each of the following numbers is a member of the sets $\mathbf{N_1}$, $\mathbf{N_0}$, $\mathbf{Z}$, $\mathbf{Q}$, $\mathbf{R}$, $\mathbf{I}$.
   (a) 3
   (b) 0
   (c) -3
   (d) 3.3

(e) 1/3
(f) π
(g) √-3

2. Say whether the following numbers are of type **Z**, **Q**, **R** or **I**.
   (a) 3
   (b) 0
   (c) -3
   (d) 3.3
   (e) 1/3
   (f) π
   (g) √-3
3. Say whether the following functions are
   - Total or partial
   - Injective

   (a) Function *mother* such that *mother(x)* gives the name of the mother of the person whose name is *x*. Assume: this function can be implemented for the whole of the human race, the argument is a string of characters, and the result is also a string of characters.

   (b) Function *not* which takes a Boolean value *true* or *false* and produces its inverse.

   (c) The *sine* function (takes a real number of radians, gives a real number).

   (d) The *tangent* function (argument and result as for *sine*).

   (e) Function *daughter* such that *daughter(x)* gives the name of the daughter of the person whose name is *x*. Same assumptions as for *mother*.

   (f) The function *divide* whose argument is a pair of real numbers, and whose result is a real number, so, for example, *divide*(3.0,2.0) = 1.5.

   (g) The function *greater_than* whose argument is a pair of real numbers, and whose result is a Boolean value (*true* or *false*), so, for example, *greater_than*(4.5,4.35) = *true*.

   (h) The function *add_window* whose argument is a pair consisting of a terminal screen and a window, and whose result is a new terminal screen which contains the window.
4. Write functions to the following specifications. The table of standard functions and operators in Appendix 1 may be helpful:

   (a) A function *neg* which returns the negative value of a number, regardless of whether a positive or negative value was given, so *neg 3 = ~3, neg ~3 = ~3.* Note that unary minus is ~ in ML.

   (b) A function which returns the (positive) fourth root of a real number, so *fourth_root 16.0 = 2.0.*

   (c) A function *tan* which returns the tangent of a real number. You can check this function out by using the predefined function arctan,

so *tan(arctan r) = r* (if *r* is in the range ±π/2).
(d) Given that *tan(x)* goes from positive to negative (via infinity) at the point where *x* = π/2, use the ML system and your *tan* function to determine a value for π.
(e) Write a function *next_char* which gives the next letter of the alphabet, so *next_char(a) = b*. Try out the function on various strange characters (ASCII characters, that is).

CHAPTER

THREE

OPERATORS AND TYPES

3.1 FUNCTIONAL COMPOSITION

To build programs out of functions we need to be able to feed the result of one function to another. This is called **composing** the functions. We can compose two functions f and g provided that f's result type is the same as g's argument type.

If

$f : 'a \rightarrow 'b$

$g : 'b \rightarrow 'c$ where *'a, 'b, 'c* are **type variables**, and stand for any type such as *int*, *real*, etc.

and

$x : 'a$

then

$f x : 'b$

$g (f x) : 'c$

Here are some examples using standard functions which are present in every ML system:

```
not (not true) = true
sqrt(real 4) = 2.0
sin (arctan 1.0) / cos (arctan 1.0) = 1.0
```

(approximately)

3.2 PREFIX AND POSTFIX NOTATION

So far we have been using **prefix** notation in which the function is placed *before* its argument:

```
f x
sin 0.66
not false
```

This is the norm for ML. On seeing two expressions following each other *a b*, ML will assume that *a* is a function being applied to its argument *b*. It will calculate the value of *b* first, and then apply *a* to it. This is called **strict evaluation**.

What would ML make of the following expression?

```
sqrt sin 0.5;

ML TYPE ERROR - Type unification failure
WANTED    :  real
FOUND     :  real -> real
```

ML knows that *sqrt* is a function of type *real* → *real*. It therefore attempts to evaluate *sin* and apply *sqrt* to it. It successfully evaluates *sin* as a function of type *real* → *real*, but when it tries to apply *sqrt* to this argument, it finds that it is of the wrong type. The correct way to write this expression is

```
sqrt (sin 0.5);

val it = 0.692406 : real
```

Now when ML attempts to evaluate the argument of *sqrt*, it finds a bracketed expression. Starting on its evaluation of this expression, it finds it must apply *sin* to the value 0.5. The types match perfectly, and the result is calculated (0.479426 : *real*, in fact). Now ML has the argument of *sqrt*, and can apply the *sqrt* function to it to give 0.692406 : *real.*

Prefixing is only a convention: we are used to talking about 'the sine of *x*' rather than '*x* sined'. But we do talk about '*x* squared'. What would happen if we reversed the convention and **postfixed** functions to their arguments? We would write things like

```
x f
0.66 sin
false not
0.5 sin sqrt
```

and ML would take two expressions *a b* to be argument *a* acted on by function *b*. If we did this, no brackets would be necessary in the last example, and the direction of evaluation would be left to right. But ML doesn't work this way, for better or worse.

3.3 OPERATORS

Human beings don't like to use postfix notation, despite its advantages. But prefix notation can be clumsy too. Which of the following representations of a well-known formula do you prefer?

```
sqrt(-(+(*(a,a),*(b,b)),*(*(*(2,a),b),cos C)

a a * b b * + 2 a * b * C cos * - sqrt

sqrt((a*a) + (b*b) - (2*a*b*cos C))
```

The final version uses **infix** notation, where the function is placed between its pair of arguments. Infix functions are also known as **binary operators**. Expressions using binary operators can be made even more compact by insisting that, for example, all multiplications are done before additions. This is called defining the **precedence** of the operators. ML has precedences for the standard operators built in (see Appendix 1 for details), which means that we can write the final version as

```
sqrt(a*a + b*b - 2*a*b*cos c)
```

The remaining brackets are needed to stop ML applying *sqrt* to the first thing it sees (*a* in this case), because the application of a prefix function like *sqrt* takes precedence over everything else.

You may be wondering how ML expresses the type of a binary operator such as + or *. To find out, let's look at another binary operator, ^, which concatenates or glues together strings of characters:

```
"hip" ^ "hop";

val it = "hiphop" : string
```

We can ask ML the type of ^ by just naming it:

```
^;
ML SYNTAX ERROR - Misplaced identifier
EXPECTING: <expression>  BUT FOUND:  ^
```

Because ^ is an infix operator, ML doesn't even recognise this as a valid expression. To find the type of ^ we must convert it to prefix form, which we can do by saying:

```
op^;

val it = fn : string * string -> string
```

The pair of strings has type *string* * *string*. In general, the notation *'a* * *'b* represents a pair of the types *'a* and *'b*. Examples:

```
(2,3)  : int * int
(true,2.0)  : bool  * real
("what",4)  : string * int
```

You may think this is rather confusing: haven't we already used * to represent multiplication? This is true, but this time we're using * between *types* and in this context it represents the **cartesian product** of two types. There is a kind of multiplication involved here, because if I have a pair (a,b) of type $\alpha * \beta$, then the number of possible values of (a,b) is the number of possible values of a times the number of possible values of b. This is why the * operator between types is called the cartesian product. (Needless to say, * is itself a binary operator, an operator which acts on types and produces a new type.)

The smart reader will have worked out that we could always use the prefix form of binary operators in ML, saying things like:

```
op+ (2,3);

val it = 5 : int
```

or

```
op^ ("pre","fix");

val it = "prefix" : string
```

but who wants to make life difficult?

Properties of Binary Operators

Certain binary operators of type $'a * 'a \rightarrow 'b$ give the same result if their arguments are swapped around. They are called **commutative** operators. Examples are ordinary + and ×, equality (=) and inequality (≠), and set union ($\cup$) and intersection ($\cap$). For any commutative operator $\oplus$, we have

$$a1 \oplus a2 = a2 \oplus a1 \qquad \text{for any } a1 \text{ and } a2$$

Binary operators which always produce a result of the same type as their arguments are said to be **closed** over the type. Examples are +, -, ×, ÷ over the type *real*, + ,- and × over the type *int* and so on. All these operators have type $'a * 'a \rightarrow 'a$.

Among these closed operators of type $'a * 'a \rightarrow 'a$, there are some that are **associative**, that is, they have the property that

$$(a \oplus b) \oplus c = a \oplus (b \oplus c) = a \oplus b \oplus c$$

This is a very convenient property, firstly because it reduces the number of brackets needed in expressions containing these operators, and secondly because such expressions can be evaluated in any order. Examples are + and × again, and also $\cup$ and $\cap$.

Finally, there is an interesting property involving pairs of closed operators. For two binary operators $\oplus$ and $\otimes$, if it is true that

$$a \oplus (b \otimes c) = (a \oplus b) \otimes (a \oplus c)$$

then we say that $\oplus$ **distributes** through $\otimes$. For example, ordinary multiplication distributes through addition. We shall find these properties of binary operators useful in the tasks that lie ahead of us.

3.4 THE COMPOSING OPERATORS

How do we compose two functions in general? Use an operator! There is no reason why a binary operator should not be applied to a pair of functions as well as to a pair of integers or reals. The result is well defined; it will be a new function which applies the two original functions (say, *f* and *g*) to the argument *x* to give *g*(*f x*). Of course, the result of applying *f* to *x* must be of the correct type to be the argument of *g*.

Historically, mathematicians have used a dot operator for function composition, saying that (g · f) x = g(f x). ML uses lower-case o, so

```
infix 3 o;
fun (g  o  f) x    =   g (f  x);

val op o = fn : ('a -> 'b) * ('c -> 'a) -> 'c -> 'b
```

Here we have specified that o is an **infix operator** of precedence 3, and then defined the operator directly. ML replies with its type, which is less than totally transparent, though it makes sense if you work on it. Part of the problem is the prefixing convention; we write *g*(*f x*) to mean apply *f* and then apply *g*. This has the consequence that long compositions of functions go backwards:

```
a o b o c o d o e o f o g
```

means apply *g* then *f* then *e* . . . and so on, ending with *a*.

An alternative is to define **forward functional composition**, to which we can give the name &. The definition is similar:

```
infix 3 &;
fun (f  &  g) x    =   g (f  x);

val op & = fn:('a -> 'b) * ('b -> 'c) -> 'a -> 'c
```

but the type is a little cleaner, and expressions now read from left to right:

```
a & b & c & d & e & f & g
```

means apply *a* then *b* then *c* . . . and so on ending with *g*.

We shall freely use both forms of functional composition as appropriate. o is a standard ML operator, but & must be defined before we can use it.

Examples

```
fun fourth_root x = (sqrt o sqrt) x;

val fourth_root = fn : real -> real
```

```
val fourth_root = sqrt o sqrt;

val fourth_root = fn : real -> real
```

```
fourth_root 16.0;

val it = 2.0 : real
```

```
fun int_sqrt n = (real & sqrt) n;

val int_sqrt = fn : int -> real
```

```
val int_sqrt = real & sqrt;

val int_sqrt = fn : int -> real
```

```
int_sqrt 4;

val it = 2.0 : real
```

These examples show another way of defining functions, directly in terms of other functions, without mentioning the argument at all. Instead of starting the definition with *fun*, which leads ML to expect a function name and an argument, we use *val*, which indicates that we are defining the value of a name. The fact that it is a function name is neither here nor there.

We now turn from operators and look at the type of thing on which they operate.

3.5 CONSTRUCTED TYPES

Types are artificial constructions that we make for our own convenience. We decree, fairly arbitrarily, that certain values shall be of a given type. Of course, some types, like *integer* or *list*, are hallowed by convention and have a great many useful and beautiful properties, but when we start to model the real world, we soon discover that we need to invent new types.

Example: A World of Fruit

Our world of fruit consists of an apple, an orange and a banana. We can model this in ML using the **datatype** declaration:

```
datatype fruit = APPLE | ORANGE | BANANA;

datatype fruit
constructor APPLE : fruit
constructor ORANGE : fruit
constructor BANANA : fruit
```

This declaration does two things: it declares the values APPLE, ORANGE, and BANANA, and coerces them to be of type *fruit*. APPLE, ORANGE and BANANA are actually **functions**; they are called **constructor functions** (or **constructors**) because they construct a value (result) from something else (argument). In this case the argument is null, and we could write APPLE as APPLE() and so on, but, as usual, we don't bother with unnecessary brackets. Using capital letters for constructor functions is a useful convention whose benefits will be apparent as we go on. (ML, unlike some computer languages, distinguishes between upper-case and lower-case letters, so a BANANA is not a banana.)

APPLE, ORANGE and BANANA are sometimes called **canonical values**, that is, values which stand for themselves and cannot be expressed in terms of anything else.

3.6 OPERATIONS ON A TYPE

Types are not there for us to sit and idly contemplate: we want to use them to obtain some useful result from our computations. We can do this by associating with each type a set of **operations**. Consider our constructed type

```
datatype fruit = APPLE | ORANGE | BANANA;
```

If we arrange our fruit in a circle and select one of the fruits, we can associate a successor fruit (result) with each selected fruit (argument) (Fig. 3.1).

We have an operation on the type *fruit* which we can express as a function *next* of type *fruit* → *fruit*.

```
val next = fn APPLE  => ORANGE
            | ORANGE => BANANA
            | BANANA => APPLE
;

val next = fn : fruit -> fruit
```

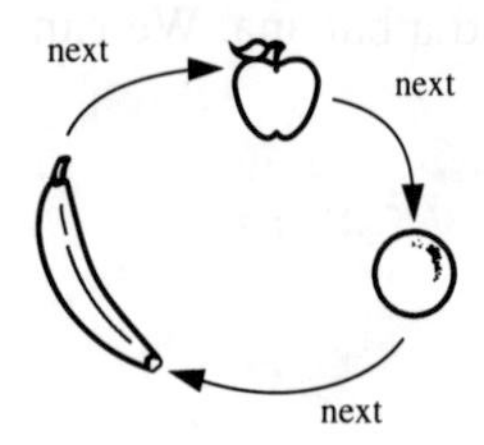

Fig. 3.1 The next fruit

Note that this is a total function because each value of *fruit* is represented on the left-hand (argument) side of the definition. It is also a surjection because each value of *fruit* is represented on the right-hand (result) side too. The symbol =>, made up of an equals sign and a greater-than sign in ML, is equivalent to the mathematical symbol $\mapsto$ which can be pronounced 'maps to'. No two values of *fruit* on the left-hand side map to the same value on the right-hand side, so we have a one-to-one function.

We can summarize these findings by saying that *next* is a total bijective function.

We can prove this function to be correct by visually checking that each argument value maps to the correct result value. This is called a **proof by exhaustion** — an appropriate name when the function has many argument values.

Exercise Try to invent a function *previous* which is the inverse of *next*, i.e. which gives the adjacent fruit on the table going in an anticlockwise direction.

Answer

```
val previous = (next o next);

val previous = fn : fruit -> fruit
```

Going clockwise twice is equivalent to going anticlockwise once!

3.7 PATTERN-MATCHING

We now come to a crucial rule about these kinds of function. Suppose ML is evaluating an expression such as

```
next BANANA
```

How will it go about it? The definition of function *next* is held as some kind of sequence of symbols in the computer store. Essentially, ML must search this sequence, find BANANA, note that it translates to ORANGE, and return ORANGE. The search could be done

- From the start of the sequence
- From the end of the sequence
- Making random leaps about the sequence
- Some other way

The rule in ML is to start from the beginning of the sequence, and continue until a match is found. This process is called **pattern-matching**. It is not a mathematical rule about functions. A mathematician, looking at the function we have just invented, would mentally translate it into Fig. 3.2.

The domain and range are the same set (our type *fruit*), and the operation *next* represents a static set of links between members of the set. We can pick out any member of the set as argument, and find its result by following the arrow. This process is assumed to be instantaneous for each member of the set.

This way of looking at the problem is fine for human beings, but the computer, in its nit-picking way, requires more guidance. Hence the rule that matching goes from left to right in the sequence of symbols representing the function. This rule has three consequences which can help us in designing functions:

- Argument values which match the early patterns in the definition of the function will be processed more quickly, in general.
- If an argument matches several patterns, only the earliest pattern will be chosen.
- The last pattern may not be strictly necessary, if it represents the only value of the type which has not yet been dealt with. In this case it can be replaced by a **wild card** symbol (_)

We can redefine *next* using a wild card:

```
val next = fn APPLE  => ORANGE
            | ORANGE => BANANA
            | _      => APPLE
;

val next = fn : fruit -> fruit
```

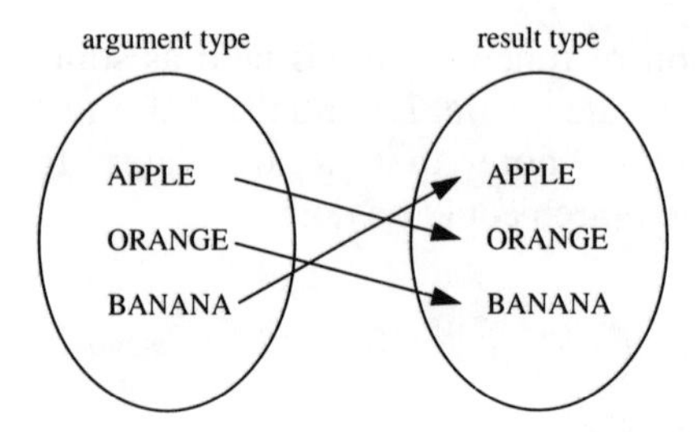

Fig. 3.2 A mathematician's view of function *next*

Redundancy and Exhaustiveness

While ML's pattern-matching rule allows us to get away with not specifying each alternative explicitly, we must be careful that our ML function definition can be translated back into the mathematician's idea of a function, namely a many-to-one mapping from a set of arguments to a set of results. So the following definition will not do:

```
val next = fn APPLE  => ORANGE
            | ORANGE => BANANA
;

ML WARNING - Rules of <match> are non-exhaustive
val next = fn : fruit -> fruit
```

Nor will this:

```
val next = fn APPLE  => ORANGE
            | ORANGE => BANANA
            | ORANGE => APPLE
            | BANANA => APPLE
;

ML WARNING - Rule 3 of <match> is redundant
val next = fn : fruit -> fruit
```

The ML type scheme, combined with the rules for pattern-matching, gives good protection against a whole range of similar errors. Although the above messages are only warnings, they should be heeded if we wish to write truly functional programs.

Incidentally, don't be fooled into believing that actual ML implementations work by laboriously searching through sequences of values in this way. In most cases, an optimization is possible. The only requirement is that the implementation should give the same result *as if* it had performed pattern matching in the way described above.

3.8 ALTERNATIVE SYNTAX FOR FUNCTIONS

There is an alternative syntax for functions which you have used already:

```
fun double i = 2 * i;
```

We can redefine *next* in this style:

```
fun next APPLE  = ORANGE
|   next ORANGE = BANANA
|   next _      = APPLE
;

val next = fn : fruit -> fruit
```

but it involves repeating the function name for each alternative, which can be tedious when there are many alternatives. Equally, we could redefine *double* using the *val* form:

```
val double = fn n => 2 * n;

val double = fn : int -> int
```

but this is a little more opaque than the *fun* form. Both forms of definition have their place in the language, and the syntax for each is given at the end of the chapter.

3.9 PROOF BY SUBSTITUTION

If a function has many possible values for its argument, proof by exhaustion can be very exhausting. Another technique for proving your program correct is **proof by substitution**. For example, show that if we invent an ML function

```
fun square (i : int) = i * i;
```

then *square* $n = n^2$ for all integer *n*.

Proof by exhaustion would require an infinite number of steps, and I do not propose to demonstrate it. But the ML definition tells us that *square i* = *i* * *i* for any value of *i* we care to choose (provided it is of type *int*), and this gives us a clue as to how we should proceed with the proof.

It is clear that we could substitute any other variable name for *i* without changing the meaning of this definition (apart from the names *square*, = and *, but I wouldn't be crazy enough to do that). (This is actually a rule of the lambda-calculus called the **alpha-conversion** rule. A variable such as *i* which can be substituted in this way is called a **bound variable**. Variables which are not bound are called **free variables**.)

So substituting *n* for *i* in this definition gives

```
square n = n * n
```

and substituting the right-hand side of this equation for the left-hand side in the proposition to be proved gives

```
n * n  = n²
```

Now, *assuming* that the ML `*` operator corresponds to normal integer multiplication, and the representation of n in ML is a valid one (i.e. that `n * n` $= n \times n$), this proposition is true by the laws of arithmetic.
QED

This is a very simple example, but it brings out some useful points about proofs in ML:

- We assume that variables in ML accurately represent concepts such as integer and real number in arithmetic. (For any implementation, this will not be strictly true — there will be some limit on the most positive and most negative integer represented, and the accuracy of representation of real numbers will also be limited. So any proof is only true in the context of these limitations.)
- We can substitute any name for the name of a bound variable, so long as we make sure that our substitutions do not cause a free variable to become bound.
- We can always substitute right-hand sides of ML definitions for left-hand sides, or left-hand sides for right-hand sides; in other words, we can substitute equals for equals.

These principles hold good for all the programs we are going to write in this book. They are incredibly powerful principles, because they make laws of algebra and arithmetic available to us in our attempts to show that our ML programs are correct. They can therefore save us many hours of testing.

Of course, the fact that we have such a powerful proof method does not preclude the need for testing altogether. We may have made a 'typo' or typographical error such as

```
fun square (i : int) = i + i;
```

and not noticed it. But once we have run a successful test (not using the value 2, in the above example!) the proof gives us a large measure of confidence in our program. Total confidence, alas, is never justified.

Another example Given our definitions

```
val next = fn APPLE  => ORANGE
            | ORANGE => BANANA
            | BANANA => APPLE
```

and

```
val previous = (next o next);
```

prove that *next* APPLE = *previous* BANANA.

Answer By substitution

previous BANANA		
=	(*next* o *next*) BANANA	definition of *previous*
=	*next* (*next* BANANA)	definition of o, alpha-conversion
=	*next* APPLE	definition of *next*

QED

3.10 SYNTAX INTRODUCED IN THIS CHAPTER

Declarations

dec	::=	`infix` ⟨ *d* ⟩ *id*	infix directive
		`val` *pat* = *exp*	value declaration
		`fun` *var* $atpat_1$ = exp_1	function declaration
		| *var* $atpat_2$ = exp_2	
		| ...	
		| *var* $atpat_m$ = exp_m	
		`datatype` *tycon* = *conbind*	datatype declaration
d	::=	a single decimal digit	
tycon	::=	*id*	
conbind	::=	*con* ⟨ | *conbind* ⟩	
id	::=	`&`	symbolic identifier

Expressions

exp	::=	`fn` *match*	function expression
match	::=	*mrule* ⟨ | *match* ⟩	
mrule	::=	*pat* `=>` *exp*	match rule

Patterns

pat	::=	*atpat*	atomic
atpat	::=	_	wild card
		scon	special constant
		var	variable
		con	constant constructor
		`(`*pat*`)`	

3.11 CHAPTER SUMMARY

Any non-trivial program is going to consist of many functions. The "glue" that binds these functions together is functional composition. ML's typing rules mean that valid functional compositions are guaranteed to produce an answer. The process of functional composition can itself be expressed as a function (two functions, in fact: o and &).

Functions which take a pair of arguments (these functions are sometimes called operators) can be prefix, postfix or infix. Infix is preferred for common operations, but requires a convention for operator precedence. Human beings seem to prefer prefix to postfix. An infix function *f* can be converted to prefix form by saying *op f*.

We can invent new types of our own using type constructors. If a constructed type has few values, the functions that operate on it can use pattern-matching, and proof by exhaustion is a possibility. ML always pattern-matches from left to right. The wild-card pattern can be used, but care must be taken that all patterns are dealt with (exhaustiveness) and none is repeated (non-redundancy). Functions with many alternatives benefit from using the `val` syntax.

Functions should always be proved correct as well as tested. For functions with large or infinite domains, proof by substitution can be used.

EXERCISES

1. Some functional languages use a different strategy from ML's **eager** evaluation of functions. Miranda™, for instance, applies a function to its argument before that argument is evaluated, and makes a note of arguments which are identical to each other, so that they are evaluated only once (**lazy evaluation**). Discuss the pros and cons of eager and lazy evaluation, using the following functions as examples:

```
fun fst(x, y) = x

val it = fst ("hello", 1/0)


fun sqr x = x * x : int

val it = sqr (4 + 6)
```

2. Try rewriting a few well-known mathematical expressions in prefix, postfix and infix notation.
3. Why is a precedence scheme not used to reduce the number of brackets needed in prefix notation?
4. Define a function *absqrt* which will return the square root of the absolute value of a number, so *absqrt*(4.0) = *absqrt*(~4.0) = 2.0. Try defining the function using o and &. Test both versions of your function.
5. Write down ML declarations for:
   (a) The type *boolean* which has constituent values TRUE and FALSE. Note that this is a different type from the standard ML type *bool*.
   (b) A type *rainbow* which contains the normal seven colours.
6. Construct an enumerated type of your own using the ML *datatype* declaration.
7. Using a type *boolean* which you have declared in Exercise 5, write functions to perform the logical operations *Not*, *And*, *Or* and *Xor* (Exclusive or). These operations are defined in the following tables:

Not p

p	TRUE	FALSE
	FALSE	TRUE

p And q

		q	
		FALSE	TRUE
p	FALSE	FALSE	FALSE
	TRUE	FALSE	TRUE

p Or q

		q	
		FALSE	TRUE
p	FALSE	FALSE	TRUE
	TRUE	TRUE	TRUE

p Xor q

		q	
		FALSE	TRUE
p	FALSE	FALSE	TRUE
	TRUE	TRUE	FALSE

Use either *val* declarations or *fun* declarations for the functions. The infix operations should have the following precedences:

And	3
Or	2
Xor	2

8. Every expression in ML has an associated type. Here are some examples:

```
true     : bool
false    : bool
3        : int
~5       : int
3.6      : real
5E2      : real   (value 5 * 10^2 = 500)
"hello": string
"c" : string
[1,2,3]: int list
["hello", "there"]  : string list
(1,2)    : int * int (pair)
(3.7, 5, "fred", true):
       real * int * string * bool (4-tuple)
sqrt: real -> real
```

```
/      : real * real -> real
not    : bool -> bool
div    : int * int -> int
+      : int * int -> int
         or  real * real -> real
<      : int * int -> bool
         or  real * real -> bool
```

Using these examples as a guide, answer the following questions.

(a) Give the types of the following expressions (some may be invalid):
  (i) `3 < 5`
  (ii) `4 / 6`
  (iii) `(5.4, ~10)`
  (iv) `[1.2, 3E4, 5.6789]`
  (v) `sqrt 4.0`
  (vi) `5 + sqrt 16.0`
  (vii) `[]`
  (viii) `["catch",22]`
  (ix) `("catch",22)`
  (x) `(sqrt, /, div)`
  (xi) `[4 + 6]`
  (xii) `not (4.0 < 8.5)`

(b) Write expressions of the following types:
  (i) *bool list*
  (ii) *(bool → bool) list*
  (iii) *(int * int → int) * (int * int → int)*
  (iv) *(bool * bool) list*
  (v) *bool * bool list*

(c) What would be the type of the following?
  (i) A function that returns the square of an integer
  (ii) A function that returns the greatest of three real numbers
  (iii) A function that takes a function *f* of type *'a -> 'b* and a *'a list* and applies *f* to each member of the list to give a *'b list.*

9. Create a datatype *day* whose canonical values are the days of the week.
10. Define functions *next_day* and *prev_day* which take an argument of type *day* and return the next day and previous day respectively.
11. Show that *next_day* = $prev\_day^{-1}$.
12. Define a function *next_day_but_one* such that e.g. *next_day_but_one MON* = *WED.*
13. Write functions that associate colours with each day of the week: *daycolour* — all days have the same colour

daycolourx — all days except Sunday have the same colour
daycolours — all days have different colours.

14. Prove that *daycolour SUN* = *daycolour MON*. Check this using the ML system.
15. Prove that *daycolourx SUN* ≠ *daycolourx MON*. Check this using the ML system.
16. Prove that *daycolours* x = *daycolours* y if and only if $x = y$. Check this using the ML system for one value of x only.
17. Write a function *colour_of_next* which returns the colour of the next day of the week to the one given as argument, the colour being defined by *daycolours*.

CHAPTER
FOUR

SIMPLE AND COMPOSITE TYPES

4.1 INTRODUCTION

In this chapter, we look at the ML standard types *int*, *real*, *bool* and *string* in more detail, and we discover how to make composite types out of simple ones. We also look at the relational operators in the context of composite types, and develop the idea of pattern-matching, introduced in Chapter 3, to give us two new kinds of expression, the if-expression and the case-expression. All this work is in preparation for writing more interesting and elaborate functions in Chapter 5.

4.2 INTEGERS

Everybody thinks they understand the integers — 'ordinary numbers' that are the acme of regularity. But they have a few odd quirks that will become apparent as we go on. Integers form the familiar sequence of values:

```
...,-2,-1,0,1,2,...
```

This sequence has some familiar properties:

1. Non-finite — the sequence is open at both ends
2. Symmetric — the sequence balances about 0

3. Ordered — any two integers can be compared using the operator ≤ (less than or equal to) to give a yes or no answer
4. Closure — the data type is **closed** under
   - Addition
   - Subtraction
   - Multiplication

By closed, we mean that the result of an operation on the integers will be an integer. Notice the deliberate omission of *division* from the list of operations. In general, dividing one integer by another will *not* give an integer as result; 5/4 is not an integer, it is a rational number.

This little quirk seems an unfortunate lapse on the part of somebody, and drives a modest coach and horses through our scheme of strong typing. Apparently we can only stay inside the integers if we eschew division. This is not acceptable for a general-purpose programming language like ML.

The solution is to invent a special kind of division which does not stray outside the type *int*. This operator, called *div*, always 'rounds down' (for a precise definition of 'rounding down' see Appendix 1), so

```
5 div 4 = 1

~3 div 6 = ~1
```

Question Does this solution work? Is the type *int* now closed under addition, subtraction, multiplication and division?

Answer Unfortunately, no. There is the small matter of division by zero. Officially, this gives a result of infinity, which is not in the type *int.* (Even though there are an infinite number of numbers in the type, infinity is not one of them!) How can we stop this latest leak in the type *int*? Let us ask ML :

```
1 div 0;


ML EXCEPTION: Div
```

In ML terms, *div* is a partial function which is not defined when its second argument is zero, so an exception is raised. Notice that the exception name is the name of the function with an initial capital letter; this is the convention for exceptions in ML.

Although exceptions are an option of last resort, there seems little alternative here, as a numerical answer would be seriously misleading and could give highly erroneous results. Even a very large integer is a long way from infinity.

If the reader has experience of other programming languages, the mention of very large integers may cause some consternation. Many languages have a built-in 'largest integer' whose value is dependent on the underlying hardware. ML has no fixed limit on integers; it is the responsibility of the implementation to define what the largest integer is, and many implementations are pretty generous in their definition. For example, the Poplog system allows any integer which can be represented as a sequence of digits in the computer store — quite a large number! You must find out for yourself what the limit for an integer is in your machine. The sequence of evaluations

```
double 1;
double it;
double it;
```

and so on will soon reveal if your implementation has been stingy in defining an integer.

Operations on Integers

Here is a list of some of the standard operations on integers:

Prefix operations (highest precedence)

`~   : int -> int`	negates a number
`abs : int -> int`	returns the absolute value of a number, e.g. `abs(~3) = 3`

Precedence 7

`div : int * int -> int`	integer division, e.g. `7 div 3 = 2`
`mod : int * int -> int`	modulo, e.g. `7 mod 3 = 1`
`* : int * int -> int`	multiplication

Precedence 6

`+ : int * int -> int`	addition
`- : int * int -> int`	subtraction

The precedence rules ensure that algebraic expressions are interpreted in the standard way so that, for example

```
3 - 4 * abs ~5;

val it = ~17 : int
```

Exceptions *Neg, Prod, Sum, Diff* will be returned by the respective functions if the result is out of range. As mentioned above, exceptions *Mod* and *Div* are returned if the divisor is zero in a *mod* or *div* operation. We shall normally use the letter *i*, *j*, *k*, *l*, *m* or *n* to indicate an integer value.

4.3 REALS

The type *real* in ML is something of a misnomer; it actually consists of rational numbers. Real numbers such as $\sqrt{2}$ are not explicitly representable as a string of digits in computer systems (because of the finite limits of store mentioned earlier), so ML has to make do with a rational approximation:

```
sqrt 2.0;

val it = 1.414214 : real
```

Most computers hold rationals in mantissa/exponent form; this format is very flexible but is of limited accuracy and has definite limits on the largest and smallest absolute values which can be accommodated. There are two consequences of this: accuracy can easily be lost when, for example, the difference of two large numbers is taken; and exceptions will arise when a result goes out of range.

A significant example of the effects of limited accuracy is given by the following:

```
sqrt 2.0 = 1.414214;

val it = false : bool
```

Equality must be used with extreme care with rationals. A better formulation of the above is

```
abs(sqrt 2.0 - 1.414214) < 1.0E~5;

val it = true : bool
```

where we have made the required accuracy explicit. The value 1.0E~5 is ML-ese for 1.0×10^{-5}.

Operations on Reals

Here is a list of some of the standard operations on reals:

Prefix operations (highest precedence)

`~ : real -> real`	negates a number
`abs : real -> real`	returns the absolute value of a number, e.g. `abs(~3.0) = 3.0`
`floor : real -> int`	returns the greatest integer which is less than the argument, e.g. `floor 2.34 = 2`, `floor ~2.34 = ~3`
`real : int -> real`	returns the real number corresponding to the argument, e.g. `real 2 = 2.0`
`sqrt : real -> real`	returns the square root of the argument, or the exception *Sqrt* if the argument is negative.
`sin : real -> real`	returns the sine of the argument considered as radians
`cos : real -> real`	returns the cosine of the argument considered as radians
`arctan : real -> real`	returns the angle in radians (in the range $\pm\pi/2$) whose tangent is the argument (i.e. tan^{-1})
`exp : real -> real`	returns e^x where x is the argument
`ln : real -> real`	returns the natural logarithm of the argument

Precedence 7

`/ : real * real -> real`	division
`* : real * real -> real`	multiplication

Precedence 6

`+ : real * real -> real`	addition
`- : real * real -> real`	subtraction

Exceptions *Neg, Abs, Floor, Sqrt, Exp, Ln, Quot, Prod, Sum, Diff* will be returned by the respective functions if the result is out of range. We shall normally use the letters x, y or z to represent a real value.

The observant reader will have noticed that some of these functions have the same name as functions defined on the integers. How can a supposedly strongly typed language like ML allow such a flagrant violation of its type rules?

The answer is an unhappy compromise. Strictly speaking, + on integers and + on reals are two completely different functions, connected only by the relation

```
abs(real (a:int +i b:int) - (real a +r real b)) < eps
```

where *eps* is some arbitrarily small value, but the use of the same sign is hallowed by tradition, so these functions are **overloaded** to operate on either type. ML can always tell by examining the types of the arguments which version of the function is meant. Nevertheless, we must work *either* in the integers *or* the reals:

```
3 + 4.0;

ML TYPE ERROR - Type unification failure
WANTED    :  int * int
FOUND     :  int * real
```

Also as a result of this compromise, we sometimes have to put an explicit type constraint on a variable, as in the *square* function of Chapter 3.

4.4 BOOLEANS

In contrast to the 'real' numbers of ML, which turned out to be not quite as we imagined them to be, Booleans are straightforward. There are only two Boolean values, *true* and *false*, which means that functions which take a Boolean expression as argument need only consider two cases:

```
fun not true = false
|   not _    = true
;

val not = fn : bool -> bool
```

not is actually a predefined function of ML, like *sqrt*, so I have redefined a standard function. This is a perfectly legitimate thing to do, but it would be considered rather anti-social if I was working in a group and defined

```
fun not true = true
|   not _    = false
;

val not = fn : bool -> bool
```

In general, redefining the standard functions is not to be recommended; you may even confuse yourself!

At this point the reader will expect me to introduce other Boolean operations such as *and* and *or*. But these operations are not implemented as standard functions in ML, for a very good reason which will be made clear later. If you wish to remedy this deficiency in ML, try Exercise 7 of Chapter 3.

We shall normally use the letters p, q or r to represent a Boolean value.

4.5 STRINGS OF CHARACTERS

Booleans have only two values, strings can have an infinite number of values. (Suppose we had collected together all the possible strings of characters; we could always invent one more string by adding the character 'a' to one of the longest strings.)

ML uses strings of characters as a means of communication with the outside world. So far in this book, we have written functions which take a number and return a number, or which take a simple constructor and return another simple constructor. But, of course, most programs are much more complex than this. In particular, the communication with the outside world tends to be fairly involved, with messages passing in either direction, and complicated synchronization conditions. Can all this be fitted into the *argument* → *function application* → *result* scheme of functional programming (the '**functional paradigm**')? I hope to show later in this book that it can, but for now be assured that strings are central to the whole enterprise.

ML lacks a type *character*, so a single character has to be represented by a string of size 1. This decision was made for reasons of implementation efficiency, but it can lead to insecurities in our programs if we're not careful. In what follows, you may notice type *string* appearing when you would expect to see type *character*. The reason for this is that the expected argument is a character, which has to be represented in ML as a string of size 1.

ML assumes an underlying alphabet of 256 characters, numbered from 0 to 255. Characters with numbers 0 to 127 correspond to the ASCII character set, which has become the standard character set for computers (a table of the printable ASCII characters is given in Appendix 2). The function *ord* : *string* → *int* gives the number of its character argument in the ASCII character set; its inverse *chr* : *int* → *string* returns the character corresponding to the numeric argument.

```
ord "A";

val it = 65 : int

chr 65;

val it = "A" : string
```

A **string constant** is a sequence, between quotes ("), of zero or more **printable characters** (those numbered between 33 and 126 in the ASCII table). The size of a string is easily determined:

```
size "hat";

val it = 3 : int
```

and strings can be concatenated together easily enough:

```
"off" ^ "side";

val it = "offside" : string
```

But how can a string be analysed into its constituent characters? ML provides no operators to do this directly, but the function *explode* converts a string into a list:

```
explode "myth";

val it = ["m", "y", "t", "h"] : string list
```

and lists can be analysed easily in ML as we shall see later. There is an (almost) inverse function (guess its name) which converts a string list to a string:

```
implode ["t","u","b","e"];

val it = "tube" : string
```

A string need not consist merely of text. The ASCII table contains control characters for most standard computer devices, and there are various 'escape conventions' which allow all kinds of exotic devices to be controlled via a string of ASCII characters.

The standard ASCII control characters have numbers in the range 0 . . . 32. ML has a convention (similar to that on many computer keyboards) which allows a control character with number n to be represented by the alphabetic character with number $n + 64$. So, for example, string `"\^A"` represents the control character SOH.

Because the backslash character (\) is used as an escape character in ML, a true backslash in a string has to be represented as `"\\"`. Other useful conventions are:

```
Quote character= "\""
New line       = "\n"
Tabulation     = "\t"
```

For non-standard character sets, or extensions of the ASCII set, any character with number 0..255 can be represented by three decimal digits *ddd*, in the form "*ddd*".

Finally, to cope with very long strings, we can use the backslash as a line terminator:

```
"how long is a piece \
\of string?";

val it = "how long is a piece of string?" :
 string
```

Operations on Strings

Here is a table of string operators:

Prefix operators (highest precedence)

`size : string -> int`	returns the length in characters of a string
`chr : int -> string`	returns the *i*th character in the ASCII character set as a 1-character string, where *i* is the argument, or exception *Chr* if *i* is not in the interval [0, 255]
`ord : string -> int`	returns the number of the first character of the argument in the ASCII character set, or exception *Ord* if the argument is the empty string
`explode : string -> string list`	returns the list of characters (as single-character strings) of which the argument consists

`implode : string list -> string`	returns the string formed by concatenating all members of the list of strings given as argument

Precedence 6

`^ : string * string -> string`	string concatenation

We shall normally use the letters *s*, *t* or *u* to represent a string value.

4.6 STANDARD CONSTRUCTED TYPES

As well as the standard simple types

```
bool      e.g. true false
int       e.g. 45
real      e.g. 45.5
string    e.g. ""   "a"   "cat"
```

ML provides two standard ways of constructing complex types from simple types: **tuples** and **lists**.

Tuples

Given two expressions *a:'a* and *b:'b* we can construct a **pair**

```
(a,b) : 'a * 'b
```

For example:

```
(1,2) : int * int
(1.0,false) : real * bool
(sqrt,"sqrt") : (real-> real)  * string
```

This idea can be generalized to a **tuple** (usually pronounced ‘tupple’) , e.g. a triple:

```
(a,b,c) : 'a * 'b * 'c
```

or a quadruple (4-tuple):

```
(a,b,c,d) : 'a * 'b * 'c * 'd
```

and so on.

This allows us to write 'functions of more than one argument' in ML, e.g.:

```
add (x,y)
distance (x,y,z)
```

add and *distance* have only one argument, but that argument is a tuple!

```
add        : real * real -> real
distance   : real * real * real -> real
```

Note the use of the cartesian product operator (*) to form the type of a tuple, and remember that this is nothing to do with multiplication on numbers. (The fact that the two operators have nothing to do with each other is the reason that the same sign can be used, in mathematics as in ML.) The cartesian product operator is an operator on types, the multiplication operator is an operator on numbers.

Special Cases of Tuple

(*a*) : *'a* is just *a* : *'a*, and corresponds to the normal practice of putting brackets around an expression.

```
() : unit
```

This is the 0-tuple or unit value. Unit is a degenerate type with just this one value. As such it would seem the quintessence of uselessness, and indeed it does not play a large role in pure functional programming. We shall avoid its use in this book, but mention that it could be used for modelling constructor functions such as

```
BANANA()
```

which can be abbreviated to

```
BANANA
```

in any case.

Example of Function Using Tuple

The following function swaps two values:

```
fun swap (a,b) = (b,a);

val swap = fn : 'a * 'b -> 'b * 'a
```

There are several interesting points about this function, simple as it is. First, the tuple on the left-hand side is a *pattern* which is matched against a value by ML as it tries to evaluate the function, while the tuple on the right-hand side is an *expression* whose value is constructed from the components already matched by ML . So

```
swap (4,"what");

val it = ("what", 4) : string * int
```

Second, the function is very general, and will happily swap any pair of values, regardless of their type. (From the function's type we can see that the types of the arguments will be swapped in the result tuple.) It is in fact a **polymorphic** function.

Third, as anyone who has experience of imperative languages will agree, this functional formulation of *swap* is a good deal easier to understand than an imperative one, which always involves an extra variable.

Lists

Tuples are useful for combining expressions of different type, but they have fixed size. A pair (2,3) is not of the same type as a triple (2,3,4). Often we wish to model a sequence of unspecified length. ML insists in this case that all the constituent types are the same, so:

```
[2,3] : int list
[2,3,4] : int list
```

These two expressions have the same type (although, of course, their values are different). We may wonder in passing why ML is so insistent on all the items in a list having the same type. Would it not be convenient to have lists of mixed type? The answer to this question becomes clearer if we think in terms of processing the items in a list dynamically, rather than statically modelling reality. A mixed list of, say, integers and strings would require some kind of test on each item before we could process it. With a list of fixed type, there is no problem: we know we can process each item in the same way. Indeed, we shall see later that ML has some very elegant methods for processing lists (and other constructed types) which depend crucially on every item in the list having the same type.

Special Cases of List

[*a*] : *'a list* is **not** the same type as *a* : *'a*. It is a list which happens to have only one item.

```
[] : 'a list
```

is the null list, also known as *nil* . Because it doesn't have any items, it can't take their type, so it is given a variable type *'a*. This can be coerced to have any particular type, when so required.

Example of Function Using Lists

Here is a (partial) function to reverse a list of three elements:

```
fun reverse [a1,a2,a3] = [a3,a2,a1];

ML WARNING - Clauses of function binding are
non-exhaustive

val reverse = fn : 'a list -> 'a list
```

We can see that lists can be used as patterns to be analysed and expressions to be constructed. Once again the function is polymorphic, and will reverse a list of any type, provided that all the elements of the list have the same type. In Chapter 8 we will see how to generalize *reverse* further, so that it operates on a list of any length.

4.7 RELATIONAL OPERATORS

There are six relational operators, some of which we have met already:

>	:	`num * num -> bool`	greater than
<	:	`num * num -> bool`	less than
>=	:	`num * num -> bool`	greater than or equal
<=	:	`num * num -> bool`	less than or equal
=	:	`''a * ''a -> bool`	equal
<>	:	`''a * ''a -> bool`	not equal

where *num* * *num* is **either** *int* * *int* **or** *real* * *real*
''*a* is an **equality type**

They have precedence 4, which is lower than arithmetic operations, so

```
1 + 2 < 5 * 4
```

is treated by ML as

```
(1 + 2)  <  (5 * 4)
```

as you would expect. The equational operators

```
=     :   ''a  * ''a -> bool
<> :   ''a  * ''a -> bool
```

are defined over a wider class of types than the others, but not over *all* types. This is why double-primed type names are used, rather than single-primed ones. The double-primed types represent **equality types**. Some types of object in ML, such as functions and exceptions, cannot be compared for equality; we say they do not *admit* equality. Of course ML could easily test whether two function *names* are identical, but this is not the same thing as testing whether two functions are equal. Suppose I write a function

```
fun double n = 2 * n;

val double = fn : int -> int
```

and you write a function

```
fun double (n:int) = n + n;

val double = fn : int -> int
```

Both these functions are called *double*, and they both double an integer, but are they actually *equal*? To avoid getting into a philosophical argument, ML refuses to equate functions. Exceptions cause similar difficulties, and so are also deemed not to admit equality. The equality types are our old friends:

bool
int
real
string

and other types constructed from these types, for example lists and tuples constructed from equality types. So . . .

bool is an equality type with typical value

```
true
```

*bool * bool* is an equality type with typical value

```
(true,false)
```

*(bool * bool) list* is an equality type with typical value

```
[(true,true),(false,false)]
```

*(bool * bool) list list* is an equality type with typical value

```
[[(true,true),(false,false)],
 [(true,false),(false,true)]]
```

and so on.

Equivalence

It is worthwhile to pause here for a moment and consider the equality relation = in a little more detail. Mathematically, = is an **equivalence relation**. It is

(i) **Reflexive** $a = a$
(ii) **Symmetric** if $a = b$ then $b = a$
(iii) **Transitive** if $a = b$ and $b = c$ then $a = c$

Also, because of the many-to-one property of functions, we can say that equality is

(iv) **substitutive** if $a = b$ then $f(a) = f(b)$

We will use these properties of the = operator in reasoning about our programs and proving them correct.

Question Is <> an equivalence relation? Is the substitution rule valid for <>? Does <> have any useful properties?

Answer <> is symmetric (if a <> b then b <> a), but it is not reflexive ($a = a$), nor is it transitive (take $a = 1$, $b = 2$, $c = 1$; then a <> b and b <> c , but $a = c$). So <> is not an equivalence relation.

The substitution rule is not true for <> either. For let $f = square$, $a = 2$, $b = \sim 2$; then a <> b but *square a* = *square b*.

So symmetry is the only useful property of <>. (Note how much easier it is to disprove a proposition (such as '<> is transitive') than to prove one. To disprove a proposition requires just *one* counter-example;

to prove a proposition requires a demonstration for *all* cases. One of the greatest services which mathematics can perform for us is to show clearly when our ideas are incorrect.)

4.8 A CASE STUDY: *abs*

Suppose we are trying to implement the absolute value function *abs* for integers; how would we do it? An explicit definition, as for our apples, oranges and bananas of the previous chapter, won't do here:

```
val abs =
  fn
  ...
  |    2 => 2
  |    1 => 1
  |    0 => 0
  |   ~1 => 1
  |   ~2 => 2
  ...
;
```

We haven't enough paper (in two directions!). What we want to do is split the argument type into two subtypes, the positive and negative integers, and treat them in two different ways, negating the negative ones and returning the positive ones unchanged. (Whether 0 is defined to be a positive or a negative integer is irrelevant here.) The relational operators we have just encountered offer a way of doing this:

```
1 >= 0;

val it = true : bool

~1 >= 0;

val it = false : bool
```

We can generalize this to the expression *n* >= 0. Now whenever our argument *n* gives a true value in this expression, we want to return *n*, and whenever *n* gives a false value, we want to return ~*n*. So the basic algorithm is clear, the only problem being how to express it in ML.

Here we have recourse to one of the fundamental principles of programming, namely **functional decomposition** (sometimes known as **divide**

and conquer.* For complicated problems (ones where we can't see the answer instantly) it is often of great benefit. The basic idea (following Julius Caesar's strategy in subjugating ancient Gaul) is to split the problem into subproblems, and tackle each subproblem separately. While we are working on one subproblem, we imagine (as no doubt did Caesar) that all the other subproblems have been solved. In this way, slowly but surely, we solve the entire problem.

We can easily invent a function that will convert the value of the expression *n* >= 0 into the result *n* or ~n. It is the function defined as follows:

```
fn true => n
|  false => ~n
;
```

(This kind of function without a name is called an **anonymous function**.)

```
ML TYPE ERROR - Unbound variable
INVOLVING:   n
```

Now ML has warned us that this function is not well-defined because *n* is free but does not have a value. We could circumvent this by making *n* part of the argument and thus binding it to the function definition:

```
fn (n,true)  => n
|  (n,false)  => ~n
;
```

```
ML TYPE ERROR - Cannot determine a type for
overloaded identifier
INVOLVING:   ~ : 'ty1 -> 'ty1
```

OK ML, we'll specify that it's integer by using an explicit type constraint:

```
fn (n,true)  => n:int
|  (n,false)  => ~n
;
```

```
val it = fn : int * bool -> int
```

Now for the second subproblem. We can also easily invent the func-

*This term is also used, in some functional language texts, for a specialized class of algorithms. I am using the term here in a more general sense.

tion which, given *n*, returns the result of the expression *n* >= 0. It is the function defined as follows:

```
fn n  =>  n >= 0;

val it = fn : int -> bool
```

The third part of the problem (like Gaul, this problem has three parts) is to put these two functions together into one function, our *abs* function. Here we want to feed the result of the second function into the first function as its argument. Unfortunately, the second function produces *n* >= 0 : *bool* and the first function requires (*n*, *n*>= 0) : *int* * *bool*. No problem! We redefine the second function:

```
fn n => (n, n>= 0);

val it = fn : int -> int * bool
```

Now we can combine the two functions to form the *abs* function by using the functional composition operator , &:

```
val abs =
   (fn n => (n, n>= 0)) &
   (fn (n,true) => n:int | (n,false) => ~n)
;

val abs = fn : int -> int
```

The brackets are necessary here so that the operands of & are unambiguously defined. Now we're fairly certain we've conquered the problem, but let's just try out a few values to make sure:

```
abs 0;

val it = 0 : int

abs ~1;

val it = 1 : int

abs 1;

val it = 1 : int
```

Here I am trying **critical values**, that is, those that lie just either side of the 'dividing line' in the expression $n >= 0$. It seems that the function works.

However, before congratulating ourselves on our programming prowess, we should consider another golden rule: **review and improve**. Are we sure that our solution is the shortest and clearest, in a word, the most **elegant**, possible? Julius Caesar was not over-concerned with elegance, but mathematicians are, and it's the mathematicians we should follow here. The reasons are threefold:

- An elegant solution takes up less space in the machine's store
- An elegant solution can also take less time to execute
- An elegant solution is easier to understand, prove, test and modify

When looked at in a critical light, our solution seems to have a lot of n's floating about to no great purpose. Perhaps we made the wrong decision in response to ML's message about n being undefined. Now is the moment to apply yet another golden rule, the principle of **abstraction**. This consists of taking a particular case, and generalizing it. It is a principle that can be carried too far, but as long as we bear in mind our vow to produce elegant solutions, we hope to avoid this danger. Now, how can we generalize the function

```
fn true => n
|  false => ~n
;
```

without supplying n as an argument? Well, looked at abstractly the function is making a choice between two values, based on the value of a Boolean argument. We could therefore write a generalized choice function that does this for *any* two values:

```
fun choice (true,x,y)  = x
|   choice (false,x,y) = y
;

val choice = fn : bool * 'a * 'a -> 'a
```

(ML reminds us that the two values must have the same type for the function to be type-correct.) Our *abs* function can now be implemented by making the *choice* function choose between n and $\sim n$ according to the value of the expression $n >= 0$:

```
fun abs n = choice(n >= 0, n, ~n);

val abs = fn : int -> int
```

(ML works out the type of the argument and result from the type of the constant 0.)

```
abs 0;

val it = 0 : int

abs ~1;

val it = 1 : int

abs 1;

val it = 1 : int
```

The new *abs* function is shorter than our previous formulation and is a lot easier to understand, too, so our goal of greater elegance has been achieved. In addition, we have a bonus — the function *choice*, which abstracts the idea of choosing from two values, allows us to formulate solutions to many similar problems. For example, the *abs* function for reals:

```
fun abs_real x = choice(x >= 0.0, x, ~x);

val abs_real = fn : real -> real
```

Another example would be the function that tells us whether a character is in upper-case. If a character is in upper-case, its number in the ASCII table must lie between that of the characters 'A' and 'Z' (the upper-case characters are contiguous in the table). The condition that the character has a number greater than or equal to that of character 'A' is given by:

```
fn c => choice(ord c >= ord "A", true, false)
```

(Yes, dear reader, I know that it is also given by

```
fn c => ord c >= ord "A"
```

— have patience.)

To modify this function to check additionally whether the character has a number less than or equal to character Z, we can substitute the condition *ord c <= ord* "Z" for *true* in the above function, giving

```
fun is_uppercase c = choice(ord c >= ord "A",
                            ord c <= ord "Z",
                            false)
;

val is_uppercase = fn : string -> bool
```

As a final example, we can easily write the function that tells us whether a number has no valid ASCII character. The condition here is that the number is less than 0 or greater than 127. Once again we proceed by considering each subcondition separately. The condition that a number is less than zero is given by

```
fn n => choice (n < 0, true, false)
```

This time we substitute for *false* to obtain

```
fun has_no_ascii_char n =
  choice (n < 0, true, n > 127);

val has_no_ascii_char = fn : int -> bool
```

4.9 THE IF-EXPRESSION

Consider the following expression:

```
choice (true, 47.8, 1.0/0.0);
```

What do you think it evaluates to? Here is ML's answer:

```
ML EXCEPTION: Quot
```

Because ML insists on evaluating all the arguments before applying the function to them, an exception is raised by the unwanted argument before it can be rejected. This feature was too strict even for the designers of ML, who incorporated the **if-expression** to avoid it. The if-expression has the following syntax:

if *bool_exp* then exp_1 else exp_2

This is pure 'syntactic suga' designed to appeal to those familiar with the if-statement of imperative languages. If we imagine the function

```
fun if_fun true = exp1
|   if_fun false = exp2
;
```

then the above if-expression is equivalent to

```
if_fun bool_exp
```

Now because we are pattern-matching on the value of the Boolean expression, and pattern-matching proceeds from left to right (top to bottom), the match rule for *false* will never be explored if the match rule for *true* is satisfied. This is the kind of 'lazy' behaviour that we want from a reasonable if-expression, and we can rewrite our awkward example as follows:

```
if true then 47.8 else 1.0/0.0;

val it = 47.8 : real
```

Although the if-expression looks like an if-statement, there are some subtle differences, which you should note if you are familiar with imperative languages:

- the if-expression is an expression like any other, and can be used anywhere an expression is legal
- the else-clause is not optional as in imperative languages, but an essential part of the if-expression's value
- exp_1 and exp_2 must have the same type

Let us now rewrite our *abs* function in terms of the if-expression:

```
fun abs n = if n >= 0 then n else ~n;

val abs = fn : int -> int

abs 0;

val it = 0 : int

abs ~1;

val it = 1 : int
```

```
abs 1;

val it = 1 : int
```

Perhaps we have lost a little space over our *choice* function, but assuredly we have gained in execution time and safety, so from now on we shall use the if-expression wherever a simple choice is required.

andalso

Our *is_uppercase* function looks like this when written in terms of the if-statement:

```
fun is_uppercase c = if ord c >= ord "A"
                     then ord c <= ord "Z"
                     else false
;

val is_uppercase = fn : string -> bool
```

This is such a common situation that a special syntactic abbreviation is used for it:

exp_1 `andalso` exp_2

which translates to

`if` exp_1 `then` exp_2 `else false`

You can see that the 'lazy' behaviour of the if-expression is retained, and exp_2 will not be evaluated if exp_1 is false. The operator *andalso* is used in ML in place of a strict Boolean *and*, as it is more efficient and safer. If a strict Boolean *and* is required, we can always write a function *and* : *bool -> bool* which will evaluate its arguments before it combines them.

Here is our *is_uppercase* function written using *andalso*:

```
fun is_uppercase c =
  ord c >= ord "A" andalso ord c <= ord "Z"
;

val is_uppercase = fn : string -> bool
```

A gain in elegance here, I think you will agree.

orelse

Our *has_no_ascii_char* function looks like this when written using an if-expression:

```
fun has_no_ascii_char n =
  if n < 0 then true else n > 127;

val has_no_ascii_char = fn : int -> bool
```

Once again, this is such a common situation that some special syntax has been developed for it:

```
exp1 orelse exp2
```

translates to

```
if exp1 then true else exp2
```

giving the following code for *has_no_ascii_char*:

```
fun has_no_ascii_char n = n < 0 orelse n > 127;

val has_no_ascii_char = fn : int -> bool
```

One of the nice features of *andalso* and *orelse* is that they can be thought of as infix operators like * and +, provided a little care is exercised. The care is needed because considered as operators they are not **commutative**, so *a andalso b* is not the same expression as *b andalso a*:

```
2 = 3 andalso 2 div 0 = 3;

val it = false : bool

2 div 0 = 3 andalso 2 = 3;

ML EXCEPTION: Div
```

On the other hand, considered as operators they *are* **associative**, so

```
(a andalso b) andalso c  = a andalso (b andalso c)
```

This means that no brackets are needed in the following function, which tells us whether a character is a punctuation mark:

```
fun is_punctuation c  =
  c = "," orelse c = "." orelse
  c = ":" orelse c = ";" orelse
  c = "?" orelse c = "!"
;

val is_punctuation = fn : string -> bool
```

Because evaluation is left to right, it makes sense to put the most likely punctuation symbols, such as comma and full stop, early in the expression.

Note the two quite different uses of the = character in the declaration of *is_punctuation*: the first occurrence of = serves to *define* the left-hand side of the declaration in terms of the right-hand side; the other occurrences of = are *relational operators*. Because ML uses the same symbol for two different purposes, you must be careful to distinguish them, as has been done in the example above by starting a new line after the 'defining ='.

Question Can you think of another elegant way to define *is_punctuation*?

Answer Go back to straightforward pattern-matching:

```
val is_punctuation =
  fn "." => true
  |  "," => true
  |  ":" => true
  |  ";" => true
  |  "?" => true
  |  "!" => true
  |  _   => false
;

val is_punctuation = fn : string -> bool
```

Both versions of *is_punctuation* are equally efficient.

4.10 THE CASE-EXPRESSION

This is a straightforward generalisation of the if-expression, which is used when a choice of more than two expressions is required. The syntax is:

```
case exp of patt1 => exp1
          | patt2 => exp2
          ...
          | pattn => expn
```

If we define

```
val case_fun = fn patt1 => exp1
                | patt2 => exp2
                ...
                | pattn => expn
;
```

then the above case expression is

```
case_fun exp
```

Example of Case-expression

We could recode our function *next* from Chapter 3:

```
fun next f =
  case f of APPLE  => ORANGE
          | ORANGE => BANANA
          | BANANA => APPLE
;

val next = fn : fruit -> fruit
```

There is no advantage to be gained in this particular example, either in clarity or brevity, but the case-expression does have its uses in functions with more complicated arguments, as we shall see later, for example in Chapter 8.

4.11 SYNTAX INTRODUCED IN THIS CHAPTER

Expressions

atexp	::=	()	0-tuple
		$(exp_1,\ldots,exp_n)$	n-tuple, $n \geq 2$
		$[exp_1,\ldots,exp_n]$	list, $n \geq 0$

exp	::=	exp_1 andalso exp_2	conjunction
		exp_1 orelse exp_2	disjunction
		if exp_1 then exp_2 else exp_3	conditional
		case *exp* of *match*	case analysis

Patterns

atpat	::=	()	0-tuple
		$(pat_1, \ldots, pat_n)$	n-tuple, $n \geq 2$
		$[pat_1, \ldots, pat_n]$	list, $n \geq 0$

Types

ty	::=	bool	
		string	
		$ty_1 * \ldots * ty_n$	tuple type, $n \geq 2$
		ty list	list type

4.12 CHAPTER SUMMARY

Reals and integers in ML are similar to their namesakes in arithmetic; ML has to be fussier about division. The normal arithmetic operations + - * and the relational operations < > <= and >= are overloaded and will work on integers or reals but not a mixture of both. The downside of overloading is that ML requires an explicit type constraint in ambiguous cases.

Booleans and strings can be defined, but strangely, there is no character type in ML. There are facilities for expressing control characters and for concatenating strings. Conversion between strings and lists is straightforward.

Simple types can be combined into tuples and lists. Tuples are of fixed length but can contain items of different types, lists are of variable length but must contain items of the same type. Tuples and lists can be used as expressions or patterns.

Equality cannot be defined on all types in ML. In particular, equality between functions cannot be defined.

Non-trivial problems require splitting into smaller ones before they can be solved: this process is known as 'divide and conquer'. In addition, we should review and refine our solution. Elegance is achieved by using appropriate abstractions.

If- and case-expressions and the operators *andalso* and *orelse* provide non-strict handling of alternatives in ML.

EXERCISES

1. Write an ML function *half* that returns its (real) argument divided by 2, and gives a real result.
2. Use *half* to write an ML function *eighth* that returns one eighth of its argument.

3. Define a general compound interest function *compound* which takes three real arguments; the first argument is the rate of interest, the second is the number of years the compound interest is applied to the principal, and the third is the principal (original amount invested). The function returns the total amount you will have after the given number of years (as a real number; unfortunately you can't spend it).
4. The function *compound* is rather tedious to use if the rate of interest is fixed. Assuming a current rate of interest of 10%, use your function *compound* to create another function *comp10* which just takes the two arguments: number of years and principal.
5. Pythagoras' theorem is well known to you. Write a function *hypotenuse* which gives the hypotenuse of a right-angled triangle, taking as argument the lengths of the other two sides. Make the type of the function *real* * *real* → *real*.
6. Write a simple digit-recognizer called *recog* which, when given a digit between 0 and 9, replies with the words 'zero', 'one', 'two', etc.
7. Define a function *even* which returns *true* if and only if its argument is an even integer.
8. Define a function *odd* which returns *true* if and only if its argument is an odd integer.
9. Define a function *greater* which returns the greater of a pair of integer arguments.
10. Define a function *greatest* which returns the greatest of a triple of integer arguments.
11. Define a function *pos* which, given an integer, returns that integer if and only if it is positive, otherwise returning zero.
12. Define a function *is_in_range*, which, given three integers, determines whether the first lies between the other two (inclusively).

CHAPTER

FIVE

REPETITION

5.1 INTRODUCTION

Most useful computer programs are repetitive in nature. As mentioned in Chapter 1, the early computer language designers abandoned the elegant and well-understood concept of recursion on grounds of efficiency, and substituted a whole range of new ways to express repetition. We have the *goto* statement and the *for* statement and the *while* statement and the *repeat* statement and so on. None of these new forms was as general or as simple as recursion, and finding ways of verifying programs containing these constructs turned out to be a non-trivial task.

Proving an imperative program correct involves learning a new mathematical language (the first-order predicate calculus), and translating each state of the program into an expression in that calculus. Understandably, most imperative programmers don't bother to prove their programs correct, relying instead on extensive testing. But it is impossible to test all the myriad combinations of states which will occur when the program is actually used. As a result, imperative programs often fail in service.

Happily, such difficulties will not trouble the reader of this book. Functional languages use one mechanism, recursion, to express every kind of repetitive process. Recursive functions can be easily proved correct using mathematical induction. No new calculus is necessary; the language of the proof is the language of the program. In fact, as we shall see, the proving process can be combined with the design of the function.

Once written, functions can be fitted together in well-understood ways so that their properties combine meaningfully (as in the *double_list* program of Chapter 1). Instead of an explosion of state-combinations, we have an orderly composition of properties.

5.2 SYNTAX OF FUNCTIONS

Before starting on recursion, let us review the syntax of functions that we have learnt so far. Here is the syntax of an ML *fun* declaration:

dec	::=	fun *function_name* pat_1 = exp_1
		| *function_name* pat_2 = exp_2
		| ...
		| *function_name* pat_m = exp_m
		;

As you can see, the formal arguments on the left-hand sides are not general expressions, but **patterns**, which the compiler can **match** against the actual argument you give the function when you want to evaluate it.

A pattern can be one of the following:

- A constant (e.g. 1)
- A constructor (e.g. ORANGE)
- A variable (e.g. *x*)
- A tuple of patterns (e.g. (*x*,*y*)), including the 0-tuple ()
- A list of patterns (e.g. [*x*,*y*,*z*]), including the null list []
- The wild card symbol _ which matches anything

An expression, on the other hand, is more general than a pattern. As well as constants, variables, tuples of expressions and lists of expressions, it can contain functions applied to arguments (e.g. tan *x*), infixed functions (e.g. $x + y$), and conditional expressions (if exp_1 then exp_2 else exp_3). In order for a function definition to be **closed**, any variable appearing in an expression on the right-hand side of a function definition must appear in the corresponding pattern on the left-hand side. We shall only write closed function definitions, so we shall stick to this rule. Another rule insisted on by ML is that any variable can appear only once in a pattern. The same variable can appear many times in an expression, however.

```
fun same (a,a)  = true
|   same (_,_)  = false
;
```

```
ML SYNTAX ERROR - Variable rebound within
declaration
INVOLVING:  a
```

```
fun duplicate a = (a,a);

val duplicate = fn : 'a -> 'a * 'a
```

This is a closed definition; the variable *a* is **local** to the definition.

```
duplicate 47;

val it = (47, 47) : int * int

duplicate "me";

val it = ("me", "me") : string * string
```

ML bans duplicate variables in patterns because variable names effectively label or identify a portion of the pattern. In the definition of function *same* above, two portions of the pattern are labelled with the same variable name. A variable name in a pattern is just a place-holder for the expression which will replace it when the function is applied, and the pattern matched. We say that the variable is **bound** to the value of the expression. If we tried to apply function *same* above, as in the expression

same (2,4)

we have a choice of two values to bind the variable *a* — an ambiguous situation.

Question How can we write a version of *same* which does not contain a repeated variable name in a pattern?

Answer The condition we are trying to express is that the first item of the pair has the same value as the second item. We can label the first item by *a*1 and the second by *a*2. Then the condition for equality is simply that *a*1 = *a*2:

```
fun same (a1,a2) = a1 = a2;

val same = fn : ''a * ''a -> bool
```

Example: the *abs* Function Revisited

We could define the *abs* function of the previous chapter, which gives the absolute value of an integer number, by cases. If the argument is positive, *abs* should return it unchanged; if, however, it is negative, *abs* should return the equivalent positive number.

```
fun abs n    = n
|   abs (~n) = n
;

ML TYPE ERROR - Unbound constructor
INVOLVING:  ~
```

ML complains that we have used a constructor function called ~ which we haven't declared. Of course, we have done no such thing; we have used the predefined negation function called ~ on the left-hand side. But ML does not expect functions in patterns, and so it takes ~ to be a constructor like ORANGE (having already ascertained that it is not a predefined constant or a variable or a tuple or a list or a wild card). Our previous definition of *abs*:

```
fun abs n = if n < 0 then ~n else n;

val abs = fn : int -> int
```

does not suffer from this problem. ML does not mind the operator ~ being used in an expression on the right-hand side.

Why is ML so insistent on this point? After all, both the function definitions for *abs* are equally valid mathematically. The answer lies in ML's pattern-matching approach to the representation of functions. Patterns without embedded operators are easier to match, so ML chooses to sacrifice generality for efficiency in this case. It is always possible to contort a function definition into a form that ML will accept, and after a while this becomes second nature to an ML programmer. Other functional languages such as Miranda™ and Haskell are less restrictive about the form of function definitions, and pay the associated price in terms of performance.

We have seen that a variable name mentioned in a pattern can appear in an expression. We know also that function names can appear in expressions. The question naturally arises: can the same function name appear on both the right-hand and the left-hand sides of a definition? The answer is that it can, and that this provides a very elegant way of expressing a repetitive process.

5.3 INDUCTION ON THE NATURAL NUMBERS

Imagine you are a promoter for a rock-concert with three bands. Your problem is to decide in which order the bands should appear at the concert. This is quite a difficult decision, as there are many ways of ordering the appearance of the bands.

Question How many ways exactly?

Answer You have three choices for the band to appear first. Then whichever band appears first, there are two bands left competing for the second position. When you have chosen the first two bands, you have exactly one band left to fit in the final slot. So altogether there are $3 \times 2 \times 1 = 6$ ways of arranging the bands.

You can see that your job gets much more difficult as the number of acts in your rock concert increases. With four bands you have $4 \times 3 \times 2 \times 1 = 24$ ways, with five bands $5 \times 4 \times 3 \times 2 \times 1 = 120$ ways of organizing things. In fact, this is our old friend the factorial function from Chapter 1. With n bands, there are $n!$ ways of arranging the concert.

The Factorial Function in ML

Let us now imagine we have never seen the factorial function, and try to design it from scratch, at the same time proving that our design is correct. How shall we express the function in ML? The straightforward pattern-matching approach suffers from the same problem as our initial definition of *abs* — a shortage of paper:

```
val factorial  =
  fn 0 => 1
  |  1 => 1
  |  2 => 2
  |  3 => 6
  |  4 => 24
  |  5 => 120
  ...
```

Note that I have defined the factorial of 0 as 1 here. This is standard mathematical practice, and corresponds to the case of the concert promoter whose bands all cancel at the last minute. The promoter has precisely one choice — to cancel the concert!

The key to finding a shorter formulation of the factorial function is the realization that there is a **recurrence relation** between successive values of the result as we gradually increase the value of the argument, so

$$\begin{aligned}
&\textit{factorial}\ 1 &= &\ 1 &&= 1 \times \textit{factorial}\ 0\\
&\textit{factorial}\ 2 &= &\ 2 \times 1 &&= 2 \times \textit{factorial}\ 1\\
&\textit{factorial}\ 3 &= &\ 3 \times 2 \times 1 &&= 3 \times \textit{factorial}\ 2\\
&\textit{factorial}\ 4 &= &\ 4 \times 3 \times 2 \times 1 &&= 4 \times \textit{factorial}\ 3\\
&\ldots
\end{aligned}$$

In general, we can say that

$$\textit{factorial}\ n = n \times \textit{factorial}\ (n\text{-}1) \qquad \text{for all } n \geq 1$$

This is a **recursive definition** of the factorial function. Note that it is not a circular definition, as in the apocryphal dictionary entry

> **recursion** *ri-kûr′shun, n.* see **recursion**

but a definition that, at each stage of its application, gets us nearer to the final result:

$$\begin{aligned}
&\textit{factorial}\ 3\\
=\ &3 \times \textit{factorial}\ 2\\
=\ &3 \times 2 \times \textit{factorial}\ 1\\
=\ &3 \times 2 \times 1 \times \textit{factorial}\ 0\\
=\ &3 \times 2 \times 1 \times 1\\
=\ &6
\end{aligned}$$

So our definition of the factorial function will consist of two cases:

- a recursive case, which represents a factorial in terms of a 'smaller' factorial, and which can be evaluated repeatedly
- a base case (*factorial* 0), which prevents the recursion from continuing indefinitely

We can now write our ML function:

```
fun factorial 0 = 1
|   factorial n = n * factorial (n-1)
;

val factorial = fn : int -> int
```

As always with pattern-matching, the order of the cases is important; the zero pattern must be placed first to be matched at all. Let's see if our function works:

```
factorial 0;

val it = 1 : int

factorial 3;

val it = 6 : int

factorial 100;

val  it  =
9332621544394415268169923885626670049071596826 4
8162146859296389521759999322991560894146397615 6
1828625369792082722375825118521091686400000000 0
00000000000000 : int
```

Moral: never try to run a rock concert with a hundred bands!

Proving the Factorial Function Correct

All the ML functions we have written so far can be proved by inspection, by exhaustion or by substitution — is there a proof technique for recursive functions? Yes, there is, and it's called **proof by induction**. In essence, we take the two cases of the function and prove them separately, then draw a rather surprising conclusion from our two proofs.

Base case We wish to prove that *factorial* 0 = 1. This is a proof by inspection of the first line of our ML definition.

Inductive case We wish to prove that *factorial* $n = n \times (n\text{-}1) \times \ldots \times 1$ for all $n \geq 0$.

Assume that *factorial* $k = k \times (k\text{-}1) \times \ldots \times 1$ for some $k \geq 0$. This is called the **inductive assumption**. Then evaluate *factorial* $(k + 1)$. $k + 1$ will always be greater than zero, so by the second line of our ML definition,

factorial $(k + 1)$
$= (k + 1) \times$ *factorial* k
(assuming ML * is identical to arithmetic $\times$)
$= (k + 1) \times k \times (k - 1) \times \ldots \times 1$ from the inductive assumption

Now for the surprising conclusion. We have shown that, for any k ($k \geq 0$), **if** *factorial* k gives the correct answer, **then** *factorial* $(k + 1)$ will also give the correct answer. And we know already that *factorial* 0 does give the correct answer. So *factorial* 1 must give the correct answer. But if *factorial* 1 gives the correct answer, then so must *factorial* 2. And if *factorial* 2 gives the correct answer, then so must *factorial* 3. And so on, for as far as we care to go in the natural numbers, in fact, for all n ($n \geq 0$):

$$factorial\ n = n \times (n\text{-}1) \times \ldots \times 1$$

QED

The Method of Differences

There is a subtle difference between the cases of our function and the cases of our proof. The function separates the cases where $n = 0$ and $n > 0$. This is necessary for the evaluation to terminate. The proof, on the other hand, deliberately overlaps the cases, so that the base case is for $n = 0$, and the inductive case is for $n \geq 0$. This is necessary for the correctness of the base case to ripple through the rest of the natural numbers.

Nevertheless, as Fig. 5.1 shows, there is a strong similarity between the structures of the function and the proof. We could say that in performing the inductive proof, we have used recursion in reverse. We have started from the base case, and proceeded upwards through the natural numbers, whereas the recursive evaluation starts from some number and proceeds downwards to zero. This similarity gives us a powerful method for creating recursive functions, called the **Method of Differences**.

When we design a recursive function, we look out for the difference between successive values of the result, as the argument increases in value. In this case, the difference was easy to spot; each time we increased the argument by one, the result was multiplied by the value of the argument. This process gives us a recurrence relation which is at the heart of both our recursive function definition and our inductive proof. The consequence is that, provided we have applied the method carefully, we can dispense with a formal inductive proof. It is prudent always to perform a mental proof on our code, though, and particularly to check that the base case and inductive case of the proof do indeed overlap.

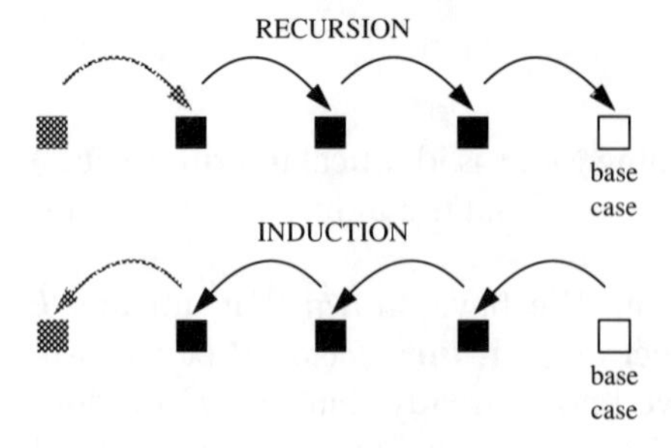

Fig. 5.1 Recursion and induction

Problems with Inductive Proofs

It only requires one link in the proof to fail for the whole proof to be invalidated. For instance suppose I decide to prove that all the natural numbers are less than 14:

1. **Base case.** $0 < 14$. True enough.
2. **Inductive case.** Assume that $k < 14$. We must show that $k + 1 < 14$. This reduces to showing that $k < 13$, which is true for all values of k except one ($k = 13$).

Because the inductive case is not quite correct, the proof fails. But it was a 'damn close-run thing'. See Exercise 3 at the end of the chapter for a subtler example of an incorrect inductive proof.

In any case, not all inductive proofs have to be complicated. As an exercise, try to prove inductively that if I wish to make one long piece of string out of n shorter pieces ($n \geq 2$), I need to tie n-1 knots.

Answers

1. **Base case.** Two pieces of string, one knot required.
2. **Inductive case.** Assume that for k pieces of string I need k-1 knots. Then when I add the $k + 1$th piece of string, I must tie another knot, making k knots.

QED

Applying the Method of Differences — Summing Numbers

Suppose we decide to write a function that will sum the numbers from 1 to n. We want to combine the proof with the design process by using the method of differences. We start by thinking about the **difference** between the result for k and the result for $k + 1$. This will be $k + 1$ and so we have our recurrence relation:

```
sum1to (k + 1)   =   (k + 1) +  sum1to  k
```

We also know the base case

```
sum1to  1   =   1
```

We can now write the function down in ML as a pattern-match on these two cases:

```
fun sum1to 1       = 1
|   sum1to (k + 1) = (k + 1) + sum1to k
;

ML TYPE ERROR - Unbound constructor
INVOLVING:   +
```

ML tells us that the expression $(k + 1)$ is not a valid pattern by the rules given at the start of the chapter. Can we modify the function definition in some way so that ML's requirements are satisfied? We recall that a valid pattern is one of the following:

- A constant (e.g. 1)
- A constructor (e.g. ORANGE)
- A variable (e.g. x)
- A tuple of patterns (e.g. (x,y)), including the 0-tuple ()
- A list of patterns (e.g. [x,y,z]), including the null list []
- The wild card symbol _ which matches anything

Let's substitute a single variable, say j, for the illegal expression $(k + 1)$. Then $k = j\text{-}1$ and we have

```
fun sum1to 1 = 1
|   sum1to j = j + sum1to (j-1)
;

val sum1to = fn : int -> int
```

Although we have proved the function, a quick test is in order:

```
sum1to 100;

val it = 5050 : int
```

How did ML arrive, apparently instantaneously, at this result?

Evaluation of Recursive Expressions

The initial expression matches the second clause in the function definition, and so evaluates as follows:

```
sum1to 100

=   100 + sum1to 99
```

Note that function application has a higher precedence than the + operator, so this expression reads 100 + (*sum1to* 99).

ML now notices that this expression contains a function application, and attempts to evaluate it further, by pattern-matching on the function definition as before:

```
100 + sum1to 99

=  100 + 99 + sum1to 98
```

Now once again there is a function application in the expression, and so ML attempts to evaluate it again. Eventually ML reaches the expression:

```
100 + 99 + 98 + ... + 3 + 2 + sum1to 1
```

ML, simple-minded as ever, tries to match the function application against the definition, and this time comes up with

```
100 + 99 + 98 + ... + 3 + 2 + 1
```

Now ML starts to evaluate the infixed + operators. As there are implicit brackets the evaluation works from right to left:

```
100 + 99 + 98 + ... + 3 + 2 + 1

=  100 + 99 + 98 + ... + 3 + 3

=  100 + 99 + 98 + ... + 6
```

and so on, finally reducing the expression to a triumphant

```
val it = 5050 : int
```

One can only marvel that so tedious a process is carried out so quickly. Note that, as usual with pattern-matching, the order of clauses is relevant:

```
fun sum1to' j  = j + sum1to' (j-1)
|   sum1to' 1 = 1
;

ML WARNING - Clause 2 of function binding is
redundant

val sum1to' = fn : int -> int
```

If we fail to heed the warning and try to evaluate an expression, disaster ensues:

```
sum1to' 100;
MISHAP - rle: RECURSION LIMIT(pop_callstack_lim)
EXCEEDED
```

Now *every* case matches the first pattern, and ML continues merrily on its way, down through zero, on through the negative numbers, finally coming to grief by using up all the computer's store with its intermediate expressions. This example shows the crucial importance of the base case: it is the basis on which we build our recursive definition; it is also the case which terminates ML's evaluation.

A Non-recursive Solution to Summing Numbers

When the German mathematician Karl Friedrich Gauss was a schoolboy, he discovered a fast way of summing a series of numbers. He was asked to add up all the numbers between 1 and 100, so he imagined the numbers laid out in a line, and then laid out a second line of the same numbers backwards. The result looked like this:

1	2	3	...	98	99	100
100	99	98	...	3	2	1

Each column added up to 101, and there were 100 columns, so the overall total was 10 100. But this, Gauss realized, is the result of adding all the numbers between 1 and 100 twice over, so the required answer is half 10 100, that is, 5050. We can imagine the teacher's surprise as young Karl put up his hand to give the answer while his classmates were toiling away with chalk and slate.

The same process will work for any series of numbers $1 \ldots n$, where n is greater than or equal to 1. In general, the columns will add up to $n + 1$, and there will be n of them, so the result will be $n \times (n + 1)/2$. As either n or $n + 1$ must be an even number, the result will always be integral.

The sceptical reader may not accept the last paragraph as a valid proof of Gauss's formula. Does the formula really work for *all* the natural numbers? Let's try to prove the formula by mathematical induction on the natural numbers. We have to show that

$$n \times (n + 1) / 2 = n + (n - 1) + \ldots + 1 \qquad \text{for all } n \geq 1$$

1. **Base case.** Substituting $n = 1$ gives $1 = 1$. True.
2. **Inductive case.** *Assume* the formula is correct for $n = k$. i.e. that

$$k \times (k + 1) / 2 = \mathrm{k} + (k - 1) + \ldots + 1$$

then we have to show that the formula is correct for $n = (k + 1)$ i.e.

$$\begin{aligned}(k+1) \times (k+2) / 2 &= (k+1) + k + (k-1) + \ldots + 1 \\ &= (k+1) + k \times (k+1) / 2 \qquad \text{(inductive assumption)} \\ &= 2 \times (k+1) / 2 + k \times (k-1)/2 \\ &= (2 \times (k+1) + k \times (k+1))/2 \\ &= (2 + k) \times (k+1)/2\end{aligned}$$

QED

As you can see, the inductive proof requires a knowledge of algebra and a willingness to perform manipulations on formulae. You may find this approach less congenial than the simple pictorial proof using columns of numbers. The use of the inductive assumption in its own proof also brings on strong attacks of disbelief in most people. But note that we are only checking *whether* the inductive assumption is justified: if it is not, then it is discarded; but if it *is* justified then the proof ripples through all the natural numbers.

Under our motto of *review and improve*, we can now use Gauss's formula to give a better method of summing numbers:

```
fun g_sum n  = n * (n+1) div 2;

val g_sum = fn : int -> int
```

Then ML has to make only *one* substitution to evaluate *g_sum n*, instead of *n* substitutions for *sum1to n*. We achieve the kind of speed-up in the computer that Gauss achieved in the classroom. But note that we still need to use induction to *prove* Gauss's formula correct.

5.4 MAKING FUNCTIONS TOTAL

Both *sum1to* and *g_sum* have a problem — they are not defined totally correctly on their argument type (*int*). For example:

- *sum1to ~1* will evaluate forever.
- *g_sum ~1* gives result 0.

Neither of these 'results' is desirable, as summing from 1 to n is only defined on the natural numbers greater than 0. There are two solutions to this problem:

1. Invent a new type *natural_number_greater_than_zero* and redefine the function on this type. Unfortunately (or fortunately) we can't do this in ML.
2. Get the system to complain if we pass an illegal integer value to the function. We can do this by using the **exception** mechanism of ML.

```
exception Safe_g_sum;

exception Safe_g_sum

fun safe_g_sum n =
  if n < 1 then raise Safe_g_sum
           else g_sum n
;

val safe_g_sum = fn : int -> int
```

As usual, the exception name is the function name with an initial capital letter. Although you may prefer something like

```
exception you_have_used_a_negative_number_you_twit;
```

the standard convention is more helpful in determining where the error occurred. This is a vital point that is often forgotten in connection with exceptions; that normally we are more concerned with *where* an error is than *what* it is. Imagine a large computer system with many hundreds of functions, all referring to each other in complex ways. Suddenly, and for no apparent reason, an exception message such as the above appears on the user's screen, and the system stops dead. The user (in addition to being insulted) has very little to go on in trying to sort things out, because the location of the problem has not been made manifest. The appearance, deep in the system, of an errant negative number may be the end result of a long series of mishaps whose causal connection is not obvious. At least the message

```
ML EXCEPTION: Safe_g_sum
```

gives the user a starting point for investigations, and, in conjunction with a careful consideration of the conditions under which the malfunction occurred, may lead to the problem being solved.

5.5 LOCAL DECLARATIONS

Although we have defined a safe version of *g_sum*, the unsafe version is still visible, and could be used by accident. To make absolutely sure that only the safe version is visible, we can define the unsafe version **locally** where nobody can get at it. A local version of a function *f* is conventionally given the name *f′*.

```
local
  fun g_sum' n  =  n * (n+1) div 2
in
  exception G_sum;
  fun g_sum n =
    if n < 1 then raise G_sum
              else g_sum' n
end;

exception G_sum
val g_sum = fn : int -> int
```

As you can see, the only values known *outside* the local declaration are the function name *g_sum* and the exception *G_sum*, these being declared between the *in* and the *end*. What, then, is the status of *g_sum′*? It is referred to *inside* the local declaration by the function *g_sum*, so it is certainly visible there. However, its **scope** does not extend beyond the word *end*, so an attempt to refer to it later is unsuccessful:

```
g_sum' 100;

ML TYPE ERROR - Unbound variable
INVOLVING:  g_sum'
```

g_sum′ is said to be a **local** function (local to the declaration of *g_sum*), whereas *g_sum* itself is a **global** function, which can be referred to anywhere after it has been declared.

There is another example of a local declaration in the above piece of ML code. The variable *n* in the definition of *g_sum′* is local to that definition, and has nothing to do with the variable *n* in the definition of *g_sum*. One test of locality is the possibility of substituting a different name for the local one. In the above example it is possible to substitute new names for both *n*'s and also *g_sum′* without changing the meaning of the declaration (provided the name *g_sum* is not used, of course!). In this sense, local variables are like the bound variables we met in Chapter 3, and we could say they are bound by the local declaration.

5.6 ANOTHER EXAMPLE: INTEGER MULTIPLICATION

Imagine we have to implement integer multiplication on a computer which only has addition and subtraction implemented in its hardware. How would we go about it? One way is to use recursion to implement an algorithm based on repeated addition. If we design the algorithm using the method of differences, we can guarantee that an inductive proof will show it to be correct.

Let us invent a function *times* which multiplies two integers together and gives an integer result. Its type will thus be $int * int \rightarrow int$. An application of the function will look like

```
times (3,4);

val it = 12 : int
```

Using the method of differences, we ask how the result of the function will change as we increment each argument. We find that

$$times(n, m{+}1) = times(n,m) + n$$

and

$$times(n{+}1, m) = times(n,m) + m$$

We choose (arbitrarily) the first of these recurrence relations to be the inductive case. The base case is pretty simple

```
times(n,1) = n
```

or better

```
times(n,0) = 0
```

because that formulation will deal with the the second number being zero (we can already deal with the first number being zero, or any other integer value).

Putting the two cases together we obtain

```
fun times (n,0) = 0
|   times (n,m) = times(n,m-1) + n
;

val times = fn : int * int -> int
```

(We've had to modify the formula a bit to make the left-hand side a pattern.) Proving this function correct for all $m \geq 0$ is now just a question of showing that

1. *times* $(n,0) = n \times 0$. This is obvious by inspection. Both sides are equal to 0.
2. If $times(n,k) = n \times k$, then $times(n,k+1) = n \times (k+1)$, for all $k \geq 0$. Substituting $k+1$ for m in the left-hand side, we obtain

$$\begin{aligned} & times(n,k+1) \\ = \;& times(n,k) + n && \text{(from the function definition)} \\ = \;& n \times k + n && \text{(from the inductive assumption)} \\ = \;& n \times k + n \times 1 && \text{(1 is the identity of multiplication)} \\ = \;& n \times (k+1) && \text{(distribution of } \times \text{ through } +) \end{aligned}$$

QED

However, typographical errors lie in wait for the unwary programmer, so a quick test seems called for:

```
times (5,6);

val it = 30 : int



times (~5,6);

val it = ~30 : int
```

and just for devilment:

```
times (5,~6);

MISHAP - rle: RECURSION LIMIT (pop_callstack_lim)
EXCEEDED
```

Because we have used an argument value outside the permitted range, the function fails to terminate at the base case. However, as the first argument can have any value, it's easy to transform any application of *times* to one where the second argument is positive, because

$$times\,(n,m) \;=\; times(\sim n,\sim m)$$

and in this way we obtain a total function:

```
local
  fun times' (n,0) = 0
  |   times' (n,m) = times'(n,m-1) + n
in
  fun times (n,m) =
    if m <0 then times'(~n,~m)
            else times'(n,m)
end
;

val times = fn : int * int -> int

times (5,6);

val it = 30 : int

times (5,0);

val it = 0 : int

times (5,~1);

val it = ~5 : int
```

We have now used a combination of inductive proof and judicious testing with critical values, and can be fairly certain that our function is correct. In practice, designing a function by the method of differences means that the inductive proof can be dispensed with, as the induction has already been done 'in reverse' as the function was created.

5.7 YET ANOTHER EXAMPLE: COUNTING DIGITS

Suppose we have been set the problem of counting the number of digits in the decimal representation of a given positive integer. We can imagine a function *digits* such that

$$\begin{aligned} \textit{digits}\ 100 &= 3 \\ \textit{digits}\ 34 &= 2 \\ \textit{digits}\ 0 &= 1 \end{aligned}$$

and so on. This seems to be a repetitive process, and yet it is hard to find a recurrence relation on the integer argument. Adding 1 to the argument *may* increase the number of digits by 1, but it may not. The key is to find a recurrence relation on the number of digits, which is also an integer, of course. A sure-fire way of increasing the number of digits by one would be to multiply the number by 10, and, conversely, dividing the number by 10 and throwing away the remainder (an operation we can represent by $\div$) will diminish the number of digits by one:

$$digits\ n = 1 + digits\ (n \div 10) \qquad (n \geq 10)$$

Proof Represent a k-digit number ($k \geq 1$) by letters $abc \ldots wxy$. We disallow leading zeroes, so a represents a digit in the range $1 \ldots 9$, while each of the other letters represents some digit in the range $0 \ldots 9$. Assume the formula is true for this number. Add another digit to the end of the number, making the $(k+1)$-digit number $abc \ldots wxyz$. If the new number represents the integer n, then $n \div 10$ is represented by the old number $abc \ldots wxy$. The minimum number of digits in $abc \ldots wxyz$ is 2, so the smallest integer represented by $abc \ldots wxyz$ is 10. The largest value represented is unlimited.
QED

This relation is a good candidate for our inductive case, especially as the operation $\div$ is represented in ML as *div*.

What is the base case for the induction? Every number must have at least one digit, so a suitable base case would be all the positive numbers that have one digit, namely the integers 0 to 9. In effect, the inductive proof is saying:

- For $0 \leq n \leq 9$, $digits\ n = 1$
- For $n \geq 10$, $digits\ n = 1 + digits\ (n \div 10)$

which is a specification of *digits* for any positive integer.

The function is built by writing down two cases in ML terms:

```
fun digits n =
  if n <= 9 then 1
  else 1 + digits (n div 10)
;

val digits = fn : int -> int
```

Remembering our motto *review and improve*, is there any way in which this function can be improved? It has been proved correct for positive integers, but you can see that negative integers will all end up in the base case! However, the number of digits in a negative number is just the same as the number of digits in its absolute value, so we can enhance the above definition:

```
local
  fun digits' n =
    if n <= 9 then 1
    else 1 + digits' (n div 10)
in
  fun digits n = digits' (abs n)
end
;

val digits = fn : int -> int
```

and test it for typical values:

```
digits 0;

val it = 1 : int

digits 9;

val it = 1 : int

digits 10;

val it = 2 : int

digits 999;

val it = 3 : int

digits ~999

val it = 3 : int
```

Once again, a combination of proof and testing has enabled us to design an efficient and robust function.

5.8 SYNTAX INTRODUCED IN THIS CHAPTER

Declarations

dec	::=	local dec_1 in dec_2 end	local declaration

Expression

exp	::=	raise *exp*	raise exception

Patterns

atpat	::=	_	wild card
		scon	special constant
		var	variable
		con	constructor
		()	0-tuple
		(*pat*)	bracketed pattern
		$(pat_1, \ldots, pat_n)$	n-tuple, $n \geq 2$
		$[pat_1, \ldots, pat_n]$	list, $n \geq 0$
pat		::=*atpat*	atomic pattern
		pat : *ty*	typed pattern

5.9 CHAPTER SUMMARY

Repetitive algorithms can be represented in functional languages by recursive function definitions, in which the function name appears on both the left- and right-hand sides. A recursive definition contains a recurrence relation which is evaluated repeatedly by the ML system. In order for the evaluation of the function to terminate, the recursion must have a well-defined base case, which is the end-point of every evaluation. Recursive functions can be proved correct using mathematical induction, and designed using the method of differences. In ML, there are restrictions on what sort of expressions can appear on the left-hand side of function definitions. The restricted expressions, called patterns, can be matched efficiently when a function application is being evaluated.

A function which fails to terminate for certain values of its argument can be made well defined by raising an exception when the illegal values are encountered. An exception should normally be given the same name as the function which raises it, and distinguished from the function name by an initial capital letter. This convention aids the detection and elimination of errors in large programs.

Local declarations allow functions which are for local consumption to be hidden from the rest of the program.

A combination of proving and testing gives confidence that our programs work as they should.

EXERCISES

1. Prove by induction that

   $$x^m \times x^n = x^{m+n}$$

   given the assumptions that $x^0 = 1$ and $x^{m+1} = x^m \times x$
   (Assume also that $\times$ is associative and commutative.)
2. Let $S_n = 1 + r + r^2 + r^3 + \ldots + r^n$. Prove by induction that $S_n = (r^{n+1} - 1)/(r - 1)$.
3. Here is a demonstration that everyone in the world has the same colour eyes. The demonstration takes the form of a proof that for any set of people, whatever the size of the set, everyone in the set has the same eye colour.
   Base case One person in the set, having two eyes (we assume), both of the same colour.
   Inductive case Assume the proposition is true for all sets of size k, where $k \geq 1$. Consider a set of size k+1, and remove any one person from the set. The remaining people form a set of size k, and therefore all have eyes of the same colour. Now remove a different person from the set of size k+1. Again, the remaining people all have eyes of the same colour. Take the union of the two sets of size k. It will be the original set of size k+1, and everyone in it will have eyes of the same colour.
   QED
   Clearly, everyone in the world does not have eyes of the same colour, and as the same argument could be used about skin-tone, religious belief, gender, etc. there must be something wrong with the argument. What is the error in the argument?
4. Use the method of differences to design a function *power* that raises an integer to a positive integer power, e.g. *power*(2,4) = 2^4 = 16, *power*(~3,3) = $(\sim 3)^3$ = ~27. Don't worry at this stage about making the function total on its second argument.
5. Prove by induction that *power* is correct for all values of its first argument and all positive values of its second argument.
6. Design a safe version of *power* called *safe_power*. It will take a pair of integers and produce an integer result or give an exception *Safe_power*.
7. Design, write and test a function *pseudo_bin* which takes a positive integer and returns its pseudo-binary representation. Pseudo-binary is a decimal number which looks like a binary number, e.g. 110 (decimal one hundred and ten) is pseudo-binary for 6. Use the Method of Differences to design a recursive function.

8. Design, write and test a function *sum_of_sq* which gives the sum of squares of numbers from 1 to n, where n is the argument. Deal sensibly with negative n.
9. Make up some recursive function of your own. Design it using the method of differences, implement it and test it.
10. In a well-known children's game, a group of people sit in a circle and shout numbers in order from 1 upwards. If the number you are about to shout contains a seven or is divisible by seven, you must shout 'peep' instead (or leave the circle!).

    Design a function *peepable* that will assist you to play this game by returning *true* if its integer argument satisfies these conditions.

    (Thanks to Åke Wikström for this excellent exercise.)

CHAPTER

SIX

HIGHER-ORDER FUNCTIONS

6.1 INTRODUCTION

Functional languages have been presented so far as an alternative to imperative languages, a different flavour of programming. A little more elegant here and there, perhaps, and a little more amenable to proof, but basically an alternative way of doing the same thing. In this chapter, we discover ways of using functional languages that have no counterpart in the imperative world. These new ideas lead to more elegant, more regular and more efficient programs, but they also change our view of what programming is all about.

In the imperative world, new features tend to be added into a language as the need arises, leading to the hotch-potch of facilities found in modern imperative languages, for example: the various kinds of loop construct, and facilities for breaking out of loops; the many different ways of structuring data; the various species of modules, packages or objects which attempt to impose a higher level of organization onto a program. The new features arrive in a steady stream, are declared indispensable by the programming establishment, and force older languages to be constantly updated, or new languages to be created, in an attempt to keep up with the latest fashion.

In 1977 John Backus, the head of the team that invented Fortran, was presented with the Turing Award, the highest honour that the computer world can bestow. Backus chose this occasion to launch an attack on conventional imperative languages, and to propose a move to functional pro-

Fig. 6.1 The von Neumann bottleneck

gramming.[1] He claimed that imperative languages were hopelessly compromised by their adherence to the von Neumann model of computing (the conventional model of a computer as a processing unit and a storage unit connected by a narrow interface — dubbed the 'von Neumann bottleneck' by Backus; see Fig. 6.1).

The obsession with moving information bit by bit across the von Neumann bottleneck was what gave imperative languages their complexity and lack of expressive power, according to Backus. In contrast, functional languages allowed the programmer to manipulate vast aggregations of data as a single unit (Fig. 6.2).

Furthermore, pointed out Backus, the arguments and results of functions can themselves be functions, allowing new 'functional forms' to be created, *without changing the syntax of the language*. These functional forms, or 'higher-order functions' are the subject of this chapter.

Now we can see how functional languages offer an escape from the 'add a feature' mentality. Because the ideas on which they are based are so fundamental and so general, new ways of programming can be accommodated without changing the language. In this way a standard dialect can emerge (as has happened with Standard ML), which may be used in very different ways for different purposes, but which, like mathematical notation, has a relatively fixed core of universally accepted conventions, which everyone can understand.

6.2 HIGHER-ORDER FUNCTIONS

The flexibility of functional languages is due to one fundamental fact: that functions are treated in the same way as other objects. Imperative languages have various ways of structuring and processing data; in functional languages these ideas are extended to functions themselves, the basic building blocks of computation. In this way the sharp distinction between

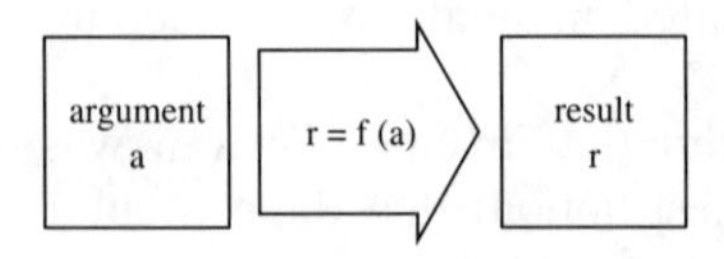

Fig. 6.2 The functional freeway

data and program (enshrined in the von Neumann model) disappears; but at the same time the strict type system of modern functional languages ensures that programs are always well defined and predictable.

Treating functions the same way as everything else means that, as Backus pointed out, functions can be arguments to other functions, and the result of a function can be a function. In other words, the argument type and result type of a function can be function types. Remember our diagrams of a function from Chapter 2? Figure 6.3 is an equivalent diagram of a higher order-function.

As you can see, the diagram is exactly the same apart from the names of the arguments and results. All the ideas that we have learnt in the previous chapters about processing data carry over, essentially unchanged, into processing functions. Also the functions that process functions can themselves be processed by functions of yet higher order.

We have met two higher-order functions already in Chapter 3; our two functional composition operators o and &. These each took a pair of functions as argument and returned a new function as result. The strict typing convention of ML meant that the argument functions could not be any old pair of functions, their types had to match in a specific way, so *sqrt* o *real* was a valid expression (representing a new function) while *real* o *sqrt* wasn't. In the following sections we will meet many more kinds of higher-order function.

6.3 CURRIED FUNCTIONS

Higher-order functions are 'old hat' to mathematicians; the ideas have been around since the 1920s, and form a well-investigated part of mathematics called **combinatory logic**, which is particularly associated with the logician Haskell B. Curry. The subject has always been one of the more theoretical parts of pure mathematics; it is only with the advent of computers and in particular functional languages that the practical application of the ideas has become possible.

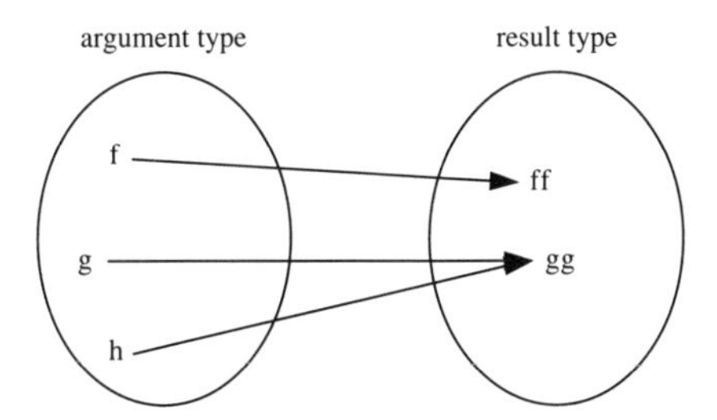

Fig. 6.3 A higher-order function

For a function which has more than one argument, we conventionally write the argument as a tuple, e.g.:

`sum (x,y)` — the sum of x and y

`radius (x,y,z)` – the distance of a point from (0,0,0)

But there is another way of looking at the matter. In the early 1920s Moses Schönfinkel, a mathematician working in Göttingen in southern Germany, became interested in the problem of reducing mathematical logic to its simplest form. By 'simple' he meant 'using the minimum number of fundamental notions'. Schönfinkel succeeded in proving that all formulae of logic could be written in terms of just three functions, which he called *C*, *S* and *U*. To achieve this result, he had to find a way of expressing functions of several arguments in terms of functions of one argument (*C*, *S* and *U* were functions of one argument).

The device he used was a simple one. Consider the function *sum* above. It is a function that takes pairs of integers and produces an integer result; in ML terms, its type is $int * int \rightarrow int$. Diagrammatically, it has the form shown in Fig. 6.4.

Schönfinkel considered an equivalent *family* of functions, *sum_1*, *sum_2*, etc., which would do the same job:

sum_1 1 = 2
sum_2 2 = 4

and so on. These functions all have type $int \rightarrow int$, and their diagrams appear as in Fig. 6.5.

So far, so simple, but Schönfinkel's next idea was revolutionary. He proposed extending the definition of a function so that a function could be an argument or result of another function. Then it was a short step to invent a function that produced the family of functions . . . ,*sum_-1*, *sum_0*, *sum_1*, . . . It would be a higher-order function, say *make_sum*, such that

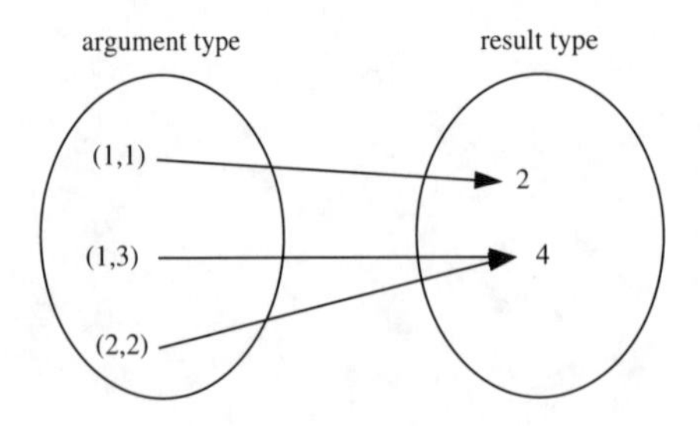

Fig. 6.4 Function sum

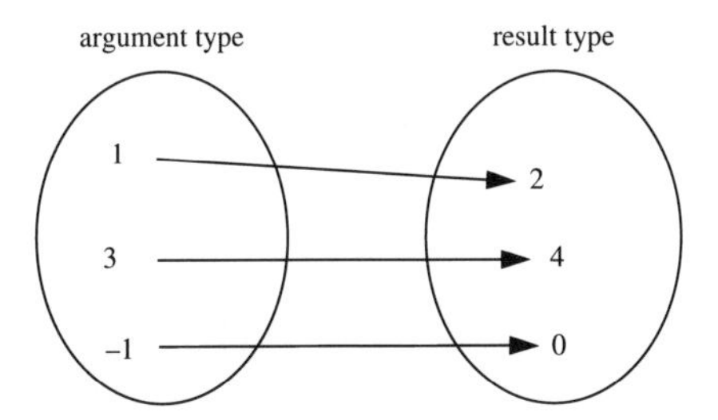

Fig. 6.5 Function *sum_1*

...
make_sum -1 = *sum_-1*
make_sum 0 = *sum_0*
make_sum 1 = *sum_1*
make_sum 2 = *sum_2*
...

having a diagram as shown in Fig. 6.6.

What is the type of this function? Well, its argument is an integer and its result is a function of type $int \rightarrow int$, so its type is clearly $int \rightarrow (int \rightarrow int)$. How could we apply the function to add two numbers, say 3 and 4? First, we would apply *make_sum* to 3 to give *sum_3*, then we would apply *sum_3* to 4 to give 7. In symbols:

make_sum 3 = *sum_3*
sum_3 4 = 7

or, removing the common variable *sum_3*:

(*make_sum* 3) 4 = 7

If we agree that function application is done from left to right, we can remove the brackets in this expression to give

make_sum 3 4 = 7

a perfectly unambiguous definition of summing 3 and 4 to give 7.

At this point the reader may be inclined to ask 'So what?' Well, the real power of this idea is its generality and simplicity — *all* the functions of more than one argument which we have encountered so far (and all those yet to come) can be redefined in this form, called a **curried function**. The curried function can then be **partially applied** to give functions like *sum_1* and so on. A great deal of clutter can be removed from

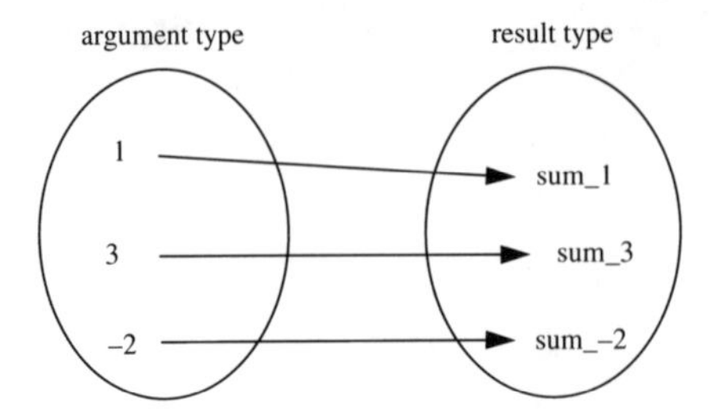

Fig. 6.6 Function *make_sum*

the notation, because all those brackets and commas associated with tuples can be dispensed with. And curried functions can themselves be used as arguments of other functions, and produced as results of other functions, giving us ways of analysing problems and constructing solutions which are just not available to imperative languages.

6.4 PARTIAL APPLICATION

Returning to our two non-curried functions:

`sum (x,y)` — the sum of x and y

`radius (x,y,z)` — the distance of a point from (0,0,0)

how can we convert them to curried form? Let's first of all invent ML versions of the functions.

```
fun sum (x,y) = x + y :int;

val sum = fn : int * int -> int
```

```
fun radius (x,y,z) = sqrt (x*x + y*y + z*z);

val radius = fn : real * real * real -> real
```

To define the curried versions of these functions could not be easier. We simply replace the uncurried pattern on the left-hand side by its curried equivalent:

```
fun make_sum x y  =  x + y :int;

val make_sum = fn : int -> int -> int

fun make_radius x y z  =  sqrt (x*x + y*y + z*z);

val make_radius = fn : real -> real -> real ->
real
```

ML gives the type of *make_sum* as *int* → *int* → *int* rather than *int* → (*int* → *int*). This means that → is taken to associate from right to left; it is a **right-associative operator**, so the two type expressions have the same meaning. By insisting that functions are applied from left to right, and that the → operator associates from right to left, we not only minimize the number of brackets in our notation but we obtain a nice correspondence between arguments and types. For instance, consider the curried function *baroque* which takes a Boolean, a real, and an integer, and returns a real number and a string. It is a pretty silly function which is only included to make a pedagogic point. Here are some results of applying this function:

baroque true 5.0 10 = (5.0,'real')

baroque false 5.0 10 = (10.0, 'integer')

The type of this function would be

bool → *real* → *integer* → *real* * *string*

because *baroque* takes a Boolean argument and produces a function of type

real → *integer* → *real* * *string*

which in turn takes a real argument to produce a function of type

integer → *real* * *string*

which in turn takes an integer argument to produce a result of type

real * *string*

Notice that the types of the curried arguments can be read off from right to left in the type expression. In general, a curried function defined as

$f\ a\ b\ c\ d \ldots = z$

will have type

$$'a \rightarrow 'b \rightarrow 'c \rightarrow 'd \rightarrow \ldots \rightarrow 'z$$

where a has type $'a$ and so on. This pleasant property follows from our decision to have function application associating to the left and the mapping operator associating to the right.

The function *baroque* also exhibits another property of curried functions; their *results* cannot be curried. All the types before the rightmost arrow represent *arguments* of the curried function, not results, and so if we want a function to *return* several values we must use a tuple. This fact limits the usefulness of currying when a function is being applied to its own result, as is sometimes necessary.

Now we have defined our *make_sum* function we can **partially apply** it. (Do not become confused between a partial function — one that is not defined for all values of its argument, and partially applying a (total) curried function. *make_sum* is a total function.) We can say

```
val inc = make_sum 1;

val inc = fn : int -> int
```

By applying *make_sum* to one argument, we have obtained a new function *inc* which increments an integer by one. We have made a more specialized function *inc*, from the more general function *make_sum*. (The *inc* function is identical to the function *sum_1* which we mentioned earlier.) As **specialization** is a useful process, which can be easily performed on curried functions, and as brackets and commas are a bother to write, we shall use curried functions from now on unless there's some burning reason not to.

Question
What would the following function do?

```
val tandoori = make_radius 43.0;

val tandoori = fn : real -> real -> real
```

Answer
tandoori y z gives the distance from the origin (0,0,0) of a point with an x coordinate of 43.0 and y- and z-coordinates y and z. In other words, it gives the distance from the origin of points in a plane given by the equa-

tion $x = 43.0$. A similar function could be easily written to give the distance of a point on the line $x = 43.0$, $y = 27.0$.

6.5 POLYMORPHIC FUNCTIONS

In chapter 4 we met the function *swap* which swapped two values in a pair:

```
fun swap (a,b) = (b,a);

val swap = fn : 'a * 'b -> 'b * 'a
```

This was a **polymorphic function** which could be applied to any pair of types. (Incidentally, note that * binds more tightly than → in this type expression.) Here is another polymorphic function:

```
fun fst (a,b) = a;

val fst = fn : 'a * 'b -> 'a
```

a and *b* can be of any type, and the result has the same type as *a* (not surprisingly), so, for instance

```
fst (2.0,6) = 2.0
```

In the expression above, *fst* has type (*real* * *int*) → *real*; in other words, the **type variables** *'a* and *'b* have the values *real* and *int* respectively. (Type variables like *'a*, *'b* and *'c* should properly be pronounced 'alpha', 'beta' and 'gamma' but most people just say 'prime-a', 'prime-b' and 'prime-c'.) Here is another example of the use of *fst*:

```
fst ((2.0,6),7) = (2.0,6)
```

This time, *fst* has type ((*real***int*) **int*) → (*real***int*), so the type variables *'a* and *'b* have values (*real***int*) and *int* respectively. Here is another example:

```
fst (fst, true) = fst
```

Question What is the type of the first *fst* in this example?

Answer In this expression, the first *fst* has type:

(*'a* * *'b* -> *'a*) * *bool* -> *'a* * *'b* -> *'a*

Let's work it out from scratch. *fst* has type *'a* * *'b* → *'a*. As the type variables in this type expression are *bound* variables, they can be replaced by any other variable provided it does not capture (i.e. match) another variable in the expression. So we could equally well say that *fst* has type *'c* * *'d* → *'c*; the meaning remains unchanged. Let's give the first *fst* this type, and obtain values for *'c* and *'d*. *'d* is obviously *bool*, and *'c* has the type of *fst*, that is, *'a* * *'b* -> *'a*. (Note that no variables have been captured by this procedure.) So the first occurrence of *first* has the type

*('a * 'b -> 'a) * bool -> 'a * 'b -> 'a*

Another question What is the value of the following expression?

```
fst (fst, true)  (2.0,6)
```

Answer 2.0

Yet another question What is the type of a type variable such as *'a* ?

Answer Type variables have no type themselves. You can imagine them to be the members of a *class* called TyVar.

The kind of polymorphism described here, where the position of an argument in the pattern determines what happens to it, irrespective of its type, is called **parametric polymorphism**. There is another kind of polymorphism which we met earlier, in Chapter 4. If we give ML the definition

```
fun double x = x + x;

ML TYPE ERROR — Cannot determine a type for
overloaded identifier
INVOLVING:  + : 'ty1 * 'ty1 -> 'ty1
```

it complains that + is an overloaded identifier. Overloading is sometimes known as ***ad hoc* polymorphism**. We have considered it to be a blemish on an otherwise unambiguously typed language. Parametric polymorphism, on the other hand, is one of the most powerful and useful features of ML, allowing us to abstract patterns of computation from the context of the particular types being computed.

6.6 THE FUNCTION *compose*

In the previous chapter we learned how to express repetition using a recursive function, and how to prove the function correct using induction. Although an inductive proof is simple when you get the hang of it, it would be pleasant if it could be bypassed for simple functions. For instance, we could imagine a function *plus* which repetitively increments a number by one to achieve its effect:

```
fun plus n 0 = n
|   plus n m = plus n (m-1) + 1
;

val plus = fn : int -> int -> int
```

The proof for *plus* would be essentially the same as that for the function *times* which we defined in Chapter 5:

```
fun times n 0 = 0
|   times n m = times n (m-1) + n
;

val times = fn : int -> int -> int
```

(This function is now curried, of course.) There is a whole class of functions called **primitive recursive functions** which have very similar proofs involving induction on the integers with zero as a base case. Because the proofs are so similar it is possible to abstract the essential recursiveness of these functions as . . . a function! Let us use the function *times* as an example. It is a primitive recursive function on two arguments, n and m. If m is zero, it returns zero. If m is greater than zero, it adds n to zero m times. In general, it adds n to zero m times, in fact, including the case where m is zero. If we were to invent a function which added n to an argument, and composed this function with itself m times, we would obtain a function which, when applied to 0, was identical to our function *times*. A table may make this clearer; let us take the case where $n = 3$. The adding function would be:

```
fun add_3 i = 3 + i;

val add_3 = fn : int -> int
```

and the multiplication table would be:

Multiplication	Using *times*	Using composition	Result
3×0	*times* 3 0	0	0
3×1	*times* 3 1	*add_3* 0	3
3×2	*times* 3 2	(*add_3* o *add_3*) 0	6
3×3	*times* 3 3	(*add_3* o *add_3* o *add_3*) 0	9
...			
$3 \times m$	*times* 3 *m*	$(add\_3)^m$ 0	$3 \times m$

Now all we need to do is design a function *compose* that will compose another function, say *f*, with itself *m* times. The inductive case is easy, we use the method of differences:

$$compose\ f\ m \quad = \quad f \circ compose\ f\ (m\text{-}1)$$

Composing a function *f* with itself *m* times is equivalent to composing it with itself *m*-1 times and then composing the result with *f*. But what about the base case? We need to return a function f^0, which represents applying *f* zero times. Fortunately there is such a function, the *identity* function mentioned in Chapter 2, which is defined as follows:

```
fun I a = a;

val I = fn : 'a -> 'a
```

The identity function just returns its argument unchanged, and that argument can be of any type.

Now we can design our *compose* function:

```
fun compose f 0 = I
|   compose f m = f o compose f (m-1)
;

val compose = fn : ('a -> 'a) -> int -> 'a -> 'a
```

Note that the function *f* must return a result of the same type as its argument. And note also that *compose* is a partial function.

With *compose* under our belt, we can easily write a function to multiply by three:

```
fun times_3 m = compose add_3 m 0;

val times_3 = fn : int -> int
```

This definition can almost be read like English; it says that the function *times_3* applied to m is the result of composing *add_3* m times and applying it to 0. Let's quickly check it out:

```
times_3 4;

val it = 12 : int
```

The last step is to generalize our *add_3* function to one which adds any two integers:

```
fun add_int n i = n + i : int;

val add_int = fn : int -> int -> int
```

Now we can produce our *times* function, by substituting the partially applied function *add_int n* for *add_3*:

```
fun times n m = compose (add_int n) m 0;

val times = fn : int -> int -> int
```

The proof of function *times* is easy, because we have encapsulated the recursion in function *compose*. We wish to show that

$$times\ n\ m \;=\; n \times m \qquad (m \geq 0)$$

Proof

Assume $add\_int\ n\ m \;=\; n + m$ — definition of *add_int*

$compose\ f\ m \;=\; f^m \;(m \geq 0)$ — definition of *compose*

then *times n m*

$= compose\ (add\_int\ n)\ m\ 0$ — definition of *times*

$= (add\_int\ n)^m\ 0$ — definition of *compose*

$= \underbrace{(n + \ldots + n)}_{(m\ \text{times})} + 0$ — definition of *add_int*

$= n \times m + 0$ — definition of $\times$

$= n \times m$ — zero is identity of addition

QED

To eliminate typographical errors, we test the function for a few typical values:

```
times 3 0;

val it = 0 : int

times 3 10;

val it = 30 : int

times ~3 10;

val it = ~30 : int

times 3 ~10;

MISHAP - rle: RECURSION LIMIT (pop_callstack_lim)
EXCEEDED
```

The behaviour on negative m is untidy because *compose* is undefined for $m < 0$. But we can easily rectify this:

```
exception Compose
fun compose f m =
  if m > 0 then f o compose f (m-1)
  else if m = 0 then I
  else raise Compose
;

exception Compose
val compose = fn : ('a -> 'a) -> int -> 'a -> 'a
```

(The cases have been put in the order of their expected likelihood.) The new version of *compose* has the following definition:

compose f m = f^m	($m \geq 0$)
compose f m raises exception *Compose*	($m < 0$)

If we redefine *times* as before:

```
fun times n m = compose (add_int n) m 0;

val times = fn : int -> int -> int
```

the proof will be the same as before, but the behaviour will be slightly different:

```
times 3 ~10;

ML EXCEPTION: Compose
```

Why did we have to redefine *times* to obtain this behaviour? The answer involves a subtle issue called **binding**. When we redefined *compose*, the new definition took effect from that point onwards, it was *not* retrospective. As *times* had been defined before we redefined *compose*, it still used the old definition of *compose*, and had to be redefined to take advantage of the new definition of *compose*. This is called **static binding**.

Should definitions be retrospective? The question has exercised legal minds for centuries, so we cannot expect to find a quick answer. There are certainly times when we would like a new definition to be applied everywhere (**dynamic binding**). The problem is that we would also like a name, once defined, to keep its value; a property that rejoices in the name of **referential transparency**. We use referential transparency every time we make a substitution in a proof, for instance. These two ideals are mutually antagonistic, so some compromise has to be reached. ML's static binding rule means that we can determine the value of a name by simply scanning the text of a program. A name retains its value in all the text following its definition, unless and until it is redefined, whereupon the new value is used in all the following text. This seems to be the simplest practical rule.

6.7 MORE USES OF *compose*

Having invented a well-defined version of *compose*, it is now an easy matter to define most primitive recursive functions. For instance, the function *plus*, defined recursively thus:

```
fun plus n 0 = n
|   plus n m = plus n (m-1) + 1
;

val plus = fn : int -> int -> int
```

can now be defined non-recursively as:

```
fun plus n m = compose (fn i => i+1) m n;

val plus = fn : int -> int -> int
```

using the anonymous form of function definition to save space.

Now here is a situation where currying brings another advantage. Suppose I redefine the above function as

```
fun plus m n = compose (fn i => i+1) m n;

val plus = fn : int -> int -> int
```

The definition must still be valid because $n + m = m + n$. But the last two arguments on each side are the same. We are saying that applying *plus* to m and then n will give exactly the same results as applying *compose* (fn i => i+1) to m and then n. So we can simplify the definition to

```
val plus = compose (fn i => i+1);

val plus = fn : int -> int -> int
```

In other words, *plus* is just a specialization of *compose*.

Suppose we wish to raise a real number to a power, we could write a function

$$ri\_power\ x\ k \;=\; x^k$$

This is a pretty obvious case of repeated composition; we are multiplying x by itself k times. When $k = 0$, the result is 1. We have

```
fun ri_power x k =
  compose (fn y => y * x) k 1.0
;

val ri_power = fn : real -> int -> real

ri_power 2.0 0;

val it = 1.0 : real

ri_power 2.0 3;

val it = 8.0 : real

ri_power 2.0 ~3;

ML EXCEPTION: Compose
```

6.8 LIMITATIONS OF *compose*

Consider our recursive function *sum1to* of Chapter 5. It was defined

```
fun sum1to 1 = 1
|   sum1to j = sum1to (j-1) + j
;

val sum1to = fn : int -> int
```

The non-recursive definition will perform repeated addition

$$0 + 1 + 2 + 3 + 4 + \ldots + j$$

and this is *not* straightforward repeated composition of the same function, but repeated composition of a function which changes slightly (in a uniform way) from one application to the next. We can obtain this effect by making it a function of *two* arguments, the second one of which represents the current value to be added. At each stage, the function will add the second argument to the first, and increment the second argument by one. In this way, the result is accumulated in the first argument:

```
fun sum_next (m, n) = (m + n, n+1);

val sum_next = fn : int * int -> int * int
```

Now we say

```
fun sum1to j = compose sum_next j (0,1);

val sum1to = fn : int -> int * int
```

This definition gives us a pair of integers as a result. Only the first integer is useful; the second integer is the value that would be added next time the function *next* was applied. We can use the polymorphic function *fst* to extract the useful value:

```
fun sum1to j = fst (compose sum_next j (0,1));

val sum1to = fn : int -> int
```

This definition looks shorter than the original recursive one. However, the crucial question is whether the proof is more complex than in the original definition. For the new definition, three proofs are required:

1. A proof that *sum_next* returns the sum of its arguments and the second argument incremented by one. This is a proof by inspection.
2. A proof that *sum1to n* returns the sum from 1 to n for $n \geq 1$. We have to demonstrate that applying *sum_next* j times ($j \geq 1$) to the pair (0,1) will give ($\Sigma^{j}_{i=1}\, i, = j+1$). This is perfectly possible, but the proof is an inductive one of rather higher complexity than the original.
3. A proof that *fst* returns the first item of a pair of items. This is a proof by inspection.

We have three proofs rather than one, and the recursive proof is more complex than before. So in this case, the original explicitly recursive formulation is superior.

Now let's consider the function *digits* that we invented in Chapter 5. You may recall that it counted the number of digits in its integer argument. It was defined as follows:

```
local
  fun digits' 0 = 0
  |   digits' n = 1 + digits' (n div 10)
in
  fun digits 0 = 1
  |   digits n = digits' (abs n)
end
;

val digits = fn : int -> int
```

Can this function be represented in terms of repeated composition? Not directly, because the induction is on the (integral) length of the argument considered as a sequence of decimal digits. So any redefinition using *compose* is going to be tortuous in the extreme, and once again the original explicit recursion is preferred.

6.9 EFFICIENCY CONSIDERATIONS

In this section, we look at ways in which repetitive algorithms can be made more efficient. This involves the use of a new formulation of the *compose* function, called *iter*. The examples use the simple types *int* and *string*.

First, let us take the problem of producing a string of n spaces ($n \geq 0$). This is a straightforward primitive recursive function, so we can use *com-*

pose. The base case ($n = 0$) returns a null string, and the string of length $k + 1$ is the string of length k with a space character concatenated to it, for all $k \geq$ o. We can express this as

```
fun spaces n = compose (fn s => s ^ " ") n "";

val spaces = fn : int -> string

spaces 3;

val it = "   " : string
```

Let us see how ML evaluated the expression *spaces* 3.

```
spaces 3

=   compose (fn s => s ^ " ") 3 ""

=   (fn s => s ^ " ") o
    compose (fn s => s ^ " ") 2 ""

=   (fn s => s ^ " ") o
    (fn s => s ^ " ") o
    compose (fn s => s ^ " ") 1 ""

=   (fn s => s ^ " ") o
    (fn s => s ^ " ") o
    (fn s => s ^ " ") o
    compose (fn s => s ^ " ") 0 ""

=   (fn s => s ^ " ") o
    (fn s => s ^ " ") o
    (fn s => s ^ " ") o
    I ""

=   (fn s => s ^ " ") o
    (fn s => s ^ " ") o
    (fn s => s ^ " ") ""

=   (fn s => s ^ " ") o
    (fn s => s ^ " ") " "

=   (fn s => s ^ " ") "  "

=   "   "
```

This does seem a rather tedious way to produce three spaces. Can we do better? It turns out that we can if we define *compose* in a different way. Here is the redefined version, called *iter*. By convention, the arguments are the other way round, but the effect of the function is the same:

```
fun iter 0 f a = a
|   iter m f a = iter (m-1) f (f a)
;

val iter = fn : int -> ('a -> 'a) -> 'a -> 'a
```

By making the argument of the function *f* an explicit argument of *iter*, we avoid all mention of the identity function *I*, but more to the point, we can express the right-hand side of the definition in terms of function *iter* only. We have avoided bringing in the compose operator o, and the long string of composed functions which that entails. In effect we have defined repeated functional composition without explicitly composing any functions. Well, the proof of the pudding is in the eating, so let's see how *iter* copes with producing a string of three spaces. The *spaces* function is redefined

```
fun spaces m = iter m (fn s => s ^ " ") "";

val spaces = fn : int -> string
```

and the evaluation goes as follows:

```
spaces 3

=  iter 3 (fn s => s ^ " ") ""

=  iter 2 (fn s => s ^ " ") " "

=  iter 1 (fn s => s ^ " ") "  "

=  iter 0 (fn s => s ^ " ") "   "

=  "   "
```

This simple change has brought about a massive reduction in the number of evaluation steps, and reduced the space required to do the evaluation. We shall therefore normally use *iter* from now on when repetition of this kind is required. *iter* is short for **iterate**, which is the Latin-derived word for repetition.

Functions like *iter*, in which all right-hand side expressions either do not mention the function or take the form of an application of the function, are called **tail recursive** functions. The name derives from the mode of evaluation; you can see that in the evaluation of *spaces* 3, at each stage *iter*'s arguments are evaluated before *iter* is applied. So the recursive evaluation of the function comes last. In the evaluation using *compose*, by contrast, even after all the recursive evaluations of *compose* have been done, there is still a lot of work to do evaluating the sequence of composed functions. This extra evaluation takes time and space.

It is clearly a good idea to make functions tail recursive whenever possible, particularly as there are very efficient ways to implement tail recursive functions on normal everyday von Neumann machines.

One final point about *iter*; it is undefined for negative *m*. But, as in the case of *compose*, we can easily redefine it to cope with all values of *m*:

```
exception Iter
fun iter m f a =
  if m > 0 then iter (m-1) f (f a)
  else if m = 0 then a
  else raise Iter
;
```

```
exception Iter
val iter = fn : int -> ('a -> 'a) -> 'a -> 'a
```

compose = iter ?

One obvious assumption in the previous section is that *compose* and *iter* actually represent the same function. They are clearly not *identical* functions (or **intentionally equal** functions, as the jargon has it) because they are defined in quite different ways. On the other hand, we are fairly certain that they *perform* the same function, that is, applying a function *f m* times. Their types are certainly consistent with this view. We have seen that substituting *iter* for *compose* in a simple example produces the same result, albeit by a different route. What we need is a demonstration that for all possible arguments, the two functions produce the same result.

We have to show that

$$compose\ f\ m\ a \;=\; iter\ m\ f\ a \qquad \text{for all } f, m \text{ and } a$$

given the following definitions:

```
exception Compose
fun compose f m =
  if m > 0 then f o compose f (m-1)
  else if m = 0 then I
  else raise Compose
;

exception Compose
val compose = fn : ('a -> 'a) -> int -> 'a -> 'a
```

```
exception Iter
fun iter m f a =
  if m > 0 then iter (m-1) f (f a)
  else if m = 0 then a
  else raise Iter
;

exception Iter
val iter = fn : int -> ('a -> 'a) -> 'a -> 'a
```

Proof As m is integer, we can split the proof into the three cases where m is less than, equal to and greater than zero.

m < 0

compose f m a raises exception *Compose*	definition of *compose*
iter m f a raises exception *Iter*	definition of *iter*

m = 0

compose f m a = *I a*	definition of *compose*
= *a*	definition of *I*
= *iter m f a*	definition of *iter*

m > 0

Assume *compose f m a* = *iter m f a* for $m = k$	$(k \geq 0)$
then *compose f* $(k+1)$ *a*	
= (*f* o *compose f k*) *a*	definition of *compose*
= *f* (*compose f k a*)	definition of o
= *f* (*iter k f a*)	inductive assumption
= *iter k f* (*f a*)	assume this for the moment
= *iter* $(k+1)$ *f a*	definition of *iter*

We can see that the proof will go through for $m > 0$ if we can show that
$f\,(iter\;k\;f\;a) \;=\; iter\;k\;f\;(f\;a)$ for all $k \geq 0$

Base case ($k = 0$)

$f\,(iter\;k\;f\;a)$	
$=\;\; f\;a$	definition of *iter*
$=\;\; iter\;k\;f\;(f\;a)$	definition of *iter*

Inductive case

assume $f\,(iter\;k\;f\;a) \;=\; iter\;k\;f\;(f\;a)$	for some $k \geq 0$
then $f\,(iter\;(k{+}1)\;f\;a)$	
$=\;\; f\,(iter\;k\;f\;(f\;a))$	definition of *iter*
$=\;\; iter\;k\;f\;(f(f\;a))$	inductive assumption
$=\;\; iter\;(k{+}1)\;f\;(f\;a)$	definition of *iter*

Our attempt at proof leaves us in an interesting situation: we have shown that $compose\;f\;m\;a = iter\;m\;f\;a$ only if $m \geq 0$. So the two functions, although similar, are not **extensionally equal**, and cannot be substituted for each other, unless we can ensure that m will never be negative when they are used. The cause of the difference, the name of the exception, is not trivial, as we shall see later.

Although the proof was relatively simple in this case, there is no general algorithm for proving extensional equality of functions (this is a consequence of Turing's proof that there is no general algorithm that will decide whether another algorithm will terminate). This is the reason why functions are not equality types in ML.

Can we 'factorize out' the final argument in the above equality? Can we state that

$$compose\;f\;m \;=\; iter\;m\;f \qquad\qquad \text{for } m \geq 0?$$

In fact the answer to this question is no, as the astute reader may have spotted, because we can produce a counter-example:

```
compose I (~1);

ML EXCEPTION: Compose

iter (~1) I;

val it = fn : 'a -> 'a
```

The exception *Iter* will not be raised until the function represented by *it* is applied to some argument. So *compose* and *iter* are equivalent only in the restricted sense proved above:

$$compose\ f\ m\ a \;=\; iter\ m\ f\ a \qquad \text{for } m \geq 0$$

6.10 MORE EXAMPLES USING *iter*

Consider the factorial function:

$$factorial\ n \;=\; n! \;=\; n \times (n\text{-}1) \times (n\text{-}2) \times \ldots \times 1$$

We defined it using explicit recursion in Chapter 5 as follows:

```
fun factorial 0 = 1
|   factorial n = n * factorial (n-1)
;

val factorial = fn : int -> int
```

and we could easily define for all *n* if we wanted to. But note that this definition is not tail recursive, and we have seen that it is tedious in evaluation. The alternative formulation using *iter* resembles the definition of *sum1to* using *compose* above. At each iteration we must multiply the current value by *j*, where *j* increases by one for each iteration. So our updating function is

```
fn(i,j)=>(i*j,j+1)

val it = fn : int * int -> int * int
```

The base case ($n = 0$) gives a result of 1, and the initial value of the multiplier is 1 (as $1! = 1 \times 1$). As in *sum1to*, only the first item of the pair is returned, so the entire factorial function becomes

```
fun factorial n =
  fst (iter n (fn(i,j)=>(i*j,j+1)) (1,1))
;

val factorial = fn : int -> int
```

A layout rule has been used here which will be helpful as our function definitions get longer. The definition will not fit on one line, so we break it at the equals sign and indent two spaces thereafter. The final semicolon is placed underneath the initial *fun* keyword.

There is no need to perform a proof on the code, as we have created it using the method of differences (applied iteratively instead of recursively), but a few tests to check our logic are in order:

```
factorial 0;

val it = 1 : int

factorial 1;

val it = 1 : int

factorial 5;

val it = 120 : int
```

Although the design process for this version of *factorial* is more complex than for the original one, and the definition is longer (33 symbols as against 18), we end up with a more efficient version of the function. Whether the efficiency gained at execution time is worth the extra design work will depend on the exact circumstances in which the function is to be used. At any rate, the ability to trade off efficiency and complexity in this way is a useful weapon in our armoury.

The *fibonacci* function illustrates a situation where optimization is essential. The Fibonacci numbers are defined as follows:

$$
\begin{aligned}
\textit{fibonacci}\ 0 &= 0\\
\textit{fibonacci}\ 1 &= 1\\
\textit{fibonacci}\ 2 &= 1\\
\textit{fibonacci}\ 3 &= 2\\
\textit{fibonacci}\ 4 &= 3\\
\textit{fibonacci}\ 5 &= 5\\
\textit{fibonacci}\ 6 &= 8\\
\textit{fibonacci}\ 7 &= 13
\end{aligned}
$$

Each Fibonacci number is obtained by adding the previous two numbers. A recursive definition seems called for:

```
fun fibonacci 0 = 0
|   fibonacci 1 = 1
|   fibonacci n = fibonacci (n-1) +
                  fibonacci (n-2)
;

val fibonacci = fn : int -> int
```

This is clearly not tail-recursive, and a sample evaluation reveals a disturbing pattern:

```
fibonacci 5
=  fibonacci 4 + fibonacci 3

=  fibonacci 3 + fibonacci 2 +
   fibonacci 2 + fibonacci 1

=  fibonacci 2 + fibonacci 1 +
   fibonacci 1 + fibonacci 0 +
   fibonacci 1 + fibonacci 0 +
   1
```

etc.

At each stage of the iteration, the number of terms in the expression doubles, until the base cases are reached. You can experience the inefficiency of *fibonacci* yourself by typing in the above definition and trying a few values on your ML system. There is a disastrous slowdown once the value of *n* reaches the low thirties.

The inefficiency of *fibonacci* is all the more surprising because hand evaluation, listing the values as I did in the table above, is an efficient iterative process. Let us now make an efficient version of *fibonacci* by using *iter*. The iterative step takes the last pair of Fibonacci integers and produces the next Fibonacci integer, then takes the new last pair and so on. We require a function from pairs of integers to pairs of integers such that

(1,1) becomes (1,2)
(1,2) becomes (2,3)
(2,3) becomes (3,5)

and so on. In general, we have

```
fn(i,j)=>(j,i+j)
```

The initial pair of numbers is (0,1), so our efficient version of *fibonacci* is

```
fun fibonacci n =
  fst (iter n (fn(i,j)=>(j,i+j)) (0,1))
;

val fibonacci = fn : int -> int
```

The proof of this function is direct from the definition: the implicit inductive proof has already been performed for *iter*. However, a few tests won't go amiss:

```
fibonacci 0;

val it = 0 : int

fibonacci 6;

val it = 8 : int

fibonacci 40;

val it = 102334155 : int
```

A modest increase in complexity has produced a major improvement in efficiency. Try evaluating *fibonacci* 40 with the original version of the function and you will see what I mean.

6.11 SYNTAX INTRODUCED IN THIS CHAPTER

Declarations

dec	::=	`fun` $var\ atpat_{11} \ldots atpat_{1n} = exp_1$	curried
		| $var\ atpat_{21} \ldots atpat_{2n} = exp_2$	function
		| ...	declaration
		| $var\ atpat_{m1} \ldots atpat_{mn} = exp_m$	
		$dec_1\ dec_2$	sequential declaration

Expressions

atexp	::=	`let` *dec* `in` *exp* `end`

6.12 CHAPTER SUMMARY

Functional programming languages treat function values like any other value. This freedom gives great regularity and power to functional languages. Also, as Schönfinkel realized 70 years ago, allowing one function to be the result of another means that functions only ever need to take one argument. Converting a function to one-argument form (or currying it, as the jargon goes) brings the advantages of cleaner notation, partial application and specialization.

If we make our functions apply from left to right, and the $\rightarrow$ operator right associative, we obtain a nice correspondence between the order of the curried arguments and the type of the curried function.

A function which takes another function as argument or returns a function as result is called a higher-order function.

Polymorphic functions can be defined in ML, such that the treatment of an argument depends only on its position in the argument pattern, and not on its type.

We can use polymorphic higher-order functions to define two closely related notions of primitive recursion; defining a primitive recursive function in terms of these functions (*compose*, *iter*) may make proving easier, or may make the function more efficient (tail recursive).

EXERCISES

1. For each of the following functions, try to work out its type:

   (a) `fun minus (m,n) = m - n:int;`
   (b) `fun times (m,n) = m * n:int;`
   (c) `fun pair_of a = (a,a);`
   (d) `fun triple_of a = (a,a,a);`
   (e) `fun swap (a,b) = (b,a);`

2. What would be the result of:
   (a) `swap ("messy",6);`
   (b) `times (3.0, 4.0);`
   (c) `pair_of ("hello");`
   (d) `triple_of(9);`
   (e) `minus ("hello","hell");`
3. Convert the functions of Exercise 1 to curried form.
4. Invent a general function *list_of* : *'a → int → 'a list* such that *list_of a n* produces a list of *n a*s.
5. Invent a similar function *string_of* : *string → int → string* such that *string_of c n* produces a string of *n c*s . Redefine *spaces* in terms of *string_of*.
6. Invent a function *char_seq* of type *string → int → string* such that *char_seq n c* gives the sequence of *n* characters in the ASCII table starting at the character *c*.
7. Convert the *is_in_range* function of Chapter 4, Exercise 12 to curried form. Tailor the function to give functions *is_decimal_digit* (number in range 0..9) and *is_binary_digit* (number is 0 or 1).

REFERENCE

1. Backus, J. W., 'Can programming be liberated from the von Neumann style? A functional style and its algebra of programs', *Communications of the ACM*, **21**, 613-41, 1978.

CHAPTER

SEVEN

CONSTRUCTOR FUNCTIONS

7.1 INTRODUCTION

So far in this book, we have used functions as units of computation — a function does something. Even higher-order functions were of this type; they were doing something to other functions. In this chapter I want to show you how functions can be used in a more static way to construct new **data types**. Or looking at it another way, I want to show that constructing a data type is also a dynamic process, which is appropriately modelled by a function.

We shall find that types have a lot in common with computations or algorithms. In particular, they can be defined recursively in a very similar way to repetitive algorithms.

We have already seen a simple example of functions being used to create a type in Chapter 3 when we defined the new type *fruit*:

```
datatype fruit = APPLE | ORANGE | BANANA;

datatype fruit
constructor APPLE : fruit
constructor ORANGE : fruit
constructor BANANA : fruit
```

ML tells us that APPLE, ORANGE and BANANA are constructors. What it actually means by this is that they are nullary functions. We could replace APPLE by APPLE() and so on, and the constructors would then be seen to have type *unit* → *fruit*. By extension, constructor functions in ML can take other types as argument (but only one argument, there are no curried constructor functions in ML). So, for instance, if we want a new type *milliseconds*, we can base it on the already-defined type *int*:

```
datatype milliseconds = MILLISECONDS of int;

datatype milliseconds
constructor MILLISECONDS : int -> milliseconds
```

MILLISECONDS is a constructor function which converts integers to our new type *milliseconds*, so 30 milliseconds is represented as *MILLI-SECONDS* 30.

7.2 DISJOINT UNION TYPES

There is no reason why constructor functions with arguments cannot be combined as our original constant constructors were, using the | symbol. For example:

```
datatype length = YARDS of real|METRES of real;

datatype length
constructor YARDS : real -> length
constructor METRES : real -> length
```

We have now combined the types *yards* and *metres* to give a new type *length*. This is a bit like forming the union of two sets, except that the members retain allegiance to their original sets, so we can always tell whence they came. For example, YARDS 3.0 and METRES 3.0 are two distinct values of type *length*.

In effect we have coerced two types to become one, and this can be of great convenience in avoiding a multiplicity of similar types, but at the same time there is no 'loss of sovereignty' — the original types can always be recovered.

Operations on Disjoint Union Types

The operations on disjoint union types will be functions, of course. But because constructor functions can appear on the left-hand side of function definitions (this is why we distinguish them from other functions by giving them capital letters) we can use pattern-matching to separate out the cases corresponding to the various component types. So **inside** the function the various components can be distinguished (by pattern-matching), but **outside** they are considered as one type. This gives a nice encapsulation of the type. For example, if we want lengths in centimetres, we can define a function to perform the conversion:

```
datatype cm = CM of real;
fun length_in_cm (METRES x)  = CM (100.0 * x)
|   length_in_cm (YARDS x)   = CM (97.6   * x)
;

datatype cm
constructor CM : real -> cm
val length_in_cm = fn : length -> cm
```

The constructor functions have been put in brackets here. Officially, this is because each pattern in a function definition must be an atomic pattern, and expressions such as METRES *x* are non-atomic patterns. Putting brackets round a pattern converts it to an atomic pattern which can be pattern-matched in a function definition. Let's see what would happen if we disobeyed this syntactic rule and left out the brackets:

```
datatype cm = CM of real;
fun length_in_cm METRES x = CM (100.0 * x)
|   length_in_cm YARDS x  = CM (97.6   * x)
;

ML TYPE ERROR - Constructor used with wrong
arity in pattern
INVOLVING:  METRES
```

The **arity** of a function is the number of arguments it takes. ML has divided the first line of the function definition into what it thinks are atomic patterns. One of these putative atomic patterns is the word METRES. It should have one argument, but seems to have none, so the error message is given.

It is not impossible to imagine a smarter version of ML, which, on seeing a constructor, and knowing its arity, would take the next atomic pattern to be an argument. However attractive this strategy might seem at first sight, it produces complications when the arguments themselves contain constructor functions. To keep pattern-matching straightforward and efficient, ML requires all patterns to be atomic in function definitions. This requirement leads to the following rules:

- Always give constructor functions capital letters, to distinguish them clearly from other functions.
- Never use ordinary (lower-case) function names in argument patterns in a function definition.
- If the constructor takes an argument in a (curried) function definition, put both the constructor and its argument in brackets to make an atomic pattern.

If you follow these rules consistently, your code will be easy to understand, and you will have fewer syntax error messages from ML to contend with.

Returning to our function *length _in_cm*, now we have defined it we can use it in expressions like any other function:

```
length_in_cm (YARDS 3.0);

val it = CM(292.8) : cm

length_in_cm (METRES 3.0);

val it = CM(300.0) : cm
```

7.3 POLYMORPHIC CONSTRUCTED TYPES

Constructed types can be even more general in ML. Suppose we are going on a journey, and we want to express the distance in a variety of units: YARDS, METRES, MILES, etc. We can define a **polymorphic** constructed type using a type variable, just like a polymorphic function.

```
datatype 'a journey = JOURNEY of 'a;

datatype 'a journey
constructor JOURNEY : 'a -> 'a journey
```

We can now express concepts such as

```
JOURNEY (YARDS 30.0);

val it = JOURNEY(YARDS(30.0)):length journey

JOURNEY (METRES 2600.0);

val it = JOURNEY(METRES(2600.0)):length journey
```

or even

```
JOURNEY (true);

val it = JOURNEY(true): bool journey
```

Clearly, this facility, while very general, has to be used with care. It is only useful for types which can have components of *any* other type. We shall meet such types in Sec. 7.5.

7.4 THE *let* EXPRESSION

Before continuing further with examples of constructor functions, we need to introduce another piece of syntax which will be extremely useful. When defining a function, we often wish to refer several times to a common subexpression. We can use a ***let* expression** to do this. As an example, consider a function which returns the roots of the quadratic equation $ax^2 + bx + c = 0$. According to a well-known result in algebra, these are given by the expression

$$\frac{-b \pm \sqrt{(b^2 - 4ac)}}{2a}$$

Our function, called *roots*, returns the two roots as a pair, given suitable values of *a*, *b* and *c* as arguments:

```
fun roots a b c =
  let
    val d = sqrt(b*b - 4.0*a*c)
    val e = 2.0*a
  in
    ((~b-d)/e, (~b+d)/e)
  end
;

val roots = fn : real -> real -> real -> real * real
```

Use of the *let* expression allows the formulae *sqrt*(*b***b* - 4.0**a***c*) and 2.0**a* to be evaluated only once. The declarations which form the head of the *let* expression are evaluated *in order* and *before* the expression which forms its body. This can be an important consideration as we shall see later when we look at performing input and output in ML. The scope of the variables *d* and *e* extends from the end of the declarations in which they are introduced to the reserved word *end*; they are said to be **local** to the let expression.

A more homely example is a function to calculate the cost of a carpet, when a 25% reduction is given for purchases of over 100 currency units:

```
fun carpet_cost (length,width,price:real) =
  let
    val amount = length * width * price
  in
    if amount <= 100.0 then amount
                                else amount * 0.75
  end
;

val carpet_cost = fn : real * real * real -> real
```

Again, space is saved in the definition and time is saved in the evaluation. This way of using the *let* expression is called **synthetic**, because many values become synthesized into one.

Analytic Use of *let*

A *let* expression can also be used to *analyse* a compound variable into its constituent parts. This is possible because the declarations in the head of the *let* expression, like any other declaration, can include an arbitrary pattern on the left-hand side. Consider the addition of two complex numbers, where each complex number is held as a pair of reals:

```
fun complex_add x y =
  let
    val (a,b) = x
    val (c,d) = y
  in
    (a+c:real,b+d:real)
  end
;

val complex_add = fn : real * real -> real *
real -> real * real
```

The *let* expression serves to analyse the complex numbers into their real components.

let and *local* — When to Use

In some programming situations it is hard to know whether to use a *let* expression or a local declaration. For example, the function *g-sum* of Chapter 5 could have been written:

```
exception G_sum;
fun g_sum n =
  let
    fun g_sum' n  =  n * (n+1) div 2
  in
    if n < 1 then raise G_sum
             else g_sum' n
  end
;

exception G_sum
val g_sum = fn : int -> int
```

Although this is legitimate ML, it is poor style for two reasons:

- There is likely to be confusion in the reader's mind between the two different uses of the identifier *n*.
- The expression in the body of the *let* expression is a long way textually from the left-hand side of the function declaration. If a *let* expression with a large number of declarations is used, this can be a serious problem.

In view of these disadvantages, my practice in this book will be to use a local declaration for local functions, and to reserve the *let* expression for the analytic and synthetic uses described above. (As with all rules, this one may be broken occasionally.) I shall also lay out *let*s and *local*s consistently, with the keywords in the same column and the constituent expressions and declarations indented two spaces relative to the keywords.

7.5 STACKS

Here then, as promised, is a type which can have components of any other type (Fig. 7.1). It represents a stack of items. The items can be of any type: papers, dishes, chairs, etc. The same rules apply to all these kinds of stack: you can only (easily) remove the top item of the stack, and new items are added to the top of the stack. So items come off in the reverse order from that in which they were added. In real life, the concept of an

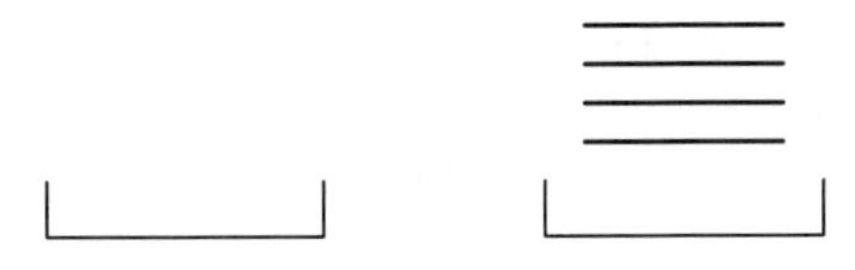

Fig. 7.1 An empty stack/a stack of four items

empty stack is not always easy to recognize (an empty in-tray, a draining board awaiting dishes, an area of floor) but for the abstract type *stack*, the concept is a vital one, because an empty stack is still of type *stack*.

The stack is a very simple abstract type, because all you can do to it is push items onto it, and pop items off it. Pushing is an operation which takes an item and a stack and produces a new stack with an extra item on the top (Fig. 7.2). If we were to model this operation as a function, what would be the type of the argument and result? The item can be of any type, so it will have a type variable (*'a*) as its type. The stack can also be of any type, but it must be of the same type as the item. (Ever tried stacking a chair on a pile of dishes? Don't.) The resulting stack will be of the same type as the original stack. So we have

$$push : 'a * 'a\ stack \rightarrow 'a\ stack$$

I have left this function in uncurried form because I want to make it an infixed operation, like + or o. We can imagine using the function to stack chairs:

```
val chairstack =
  CHAIR push (CHAIR push EMPTY_STACK);
```

The stack is building up from right to left, so if we make our push operator **right associative** (like →), we can avoid the use of brackets. The **infixr** directive specifies to ML that we are defining a right associative infixed operator. We can take the precedence of *push* to be 5, i.e. just below that of the arithmetic operators:

```
infixr 5 push;
```

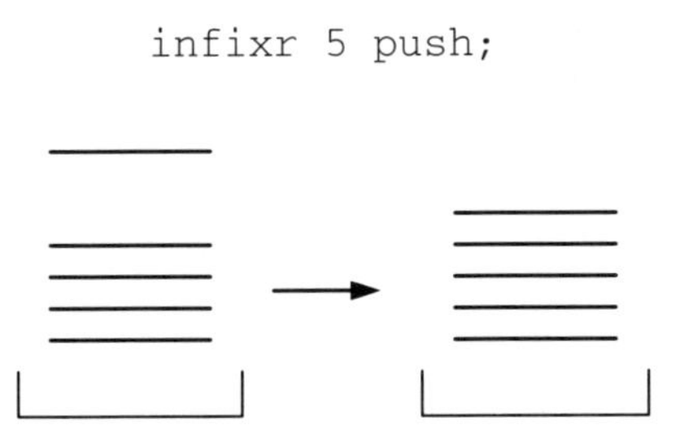

Fig. 7.2 Pushing an item onto a stack

But how do we define the function? We know the type of the two arguments, and the type of the result, but what is the mechanism for going from one to the other? Let's try a straight composition into a pair:

```
fun a push A = (a,A);

val op push = fn : 'a * 'b -> 'a * 'b
```

ML says, quite correctly, that we have defined a function that takes two arguments of type *'a* and *'b*, and combines them into a pair of type *'a * 'b*. (Note that ML refers to *op push* — this is the prefixed form of the function *push*. Whenever we use *push* in a non-infixed way, we must prefix it by *op*.) This type signature is fine as far as it goes, but does not correspond to our requirement for push, which should have type *'a * 'a stack → 'a stack*. We could redefine *push* as:

```
fun a push A = A;

val op push = fn : 'a * 'b -> 'b
```

This gives an approximation to the correct type, but at the cost of losing the semantics of the type *stack*, for it is clear that anything pushed onto this stack is lost forever. In desperation, we try to incorporate into the definition the fact that a different stack is created every time we do a *push*:

```
fun a push A = A';

ML TYPE ERROR - Unbound variable
INVOLVING: A'
```

but ML will have none of it. What are we to do? By the definition of a stack, we require that the right-hand side of the definition of push has as much information as the left-hand side; but the ML type scheme requires the result to have the same type as the second argument.

One answer to this problem is to abandon the type scheme — this is the course followed by the language Lisp, for example. We would like a less drastic solution, as the ML type scheme has proved to be of real benefit so far. Can we achieve some kind of compromise which is compatible with the rest of the language?

Let us consider a similar situation, adding two integers. The + operator has type

$$+ : int * int \rightarrow int$$

so two integers are combined into one. But there is a subtle difference — we cannot retrieve the original arguments from the result, because + is a many-to-one function. In the case of *push*, we must be able to retrieve the original arguments from the result, because the only point of stacking items is to eventually retrieve them! So *push* has to be a one-to-one function, where each distinct pair of arguments produces a distinct result. But, just as for the integers, we want a stack of any size or composition to be of type *'a stack*.

Clearly, we must coerce many types into one, and this is where ML's datatype declaration comes in. We can *define* our *push* function to be of type *'a * 'a stack → 'a stack* by making it a constructor function PUSH.

```
infixr 5 PUSH;
datatype 'a stack = op PUSH of 'a * 'a stack;

infixr 5 PUSH
datatype 'a stack
constructor op PUSH : 'a * 'a stack -> 'a stack
```

ML finds this perfectly acceptable, but now we have a logical problem. If every stack is represented as a combination of an item and a stack, then we can never arrive at the empty stack. (This is our old friend the **infinite regress** once again.) Clearly we have to bring in the empty stack somehow, as a value of the type *'a stack* (even though it contains no items of type *'a*). Once again we use the datatype declaration to coerce two types into one.

```
 infixr 5 PUSH;
 datatype 'a stack = EMPTY_STACK
                   | op PUSH of 'a * 'a stack
 ;
infixr 5 PUSH
datatype 'a stack
constructor EMPTY_STACK : 'a stack
constructor op PUSH : 'a * 'a stack -> 'a stack
```

The reader may think we are playing fast and loose with ML's type system here. But the alternative of a different type for each different size of stack is too strict to be practical, and does not correspond to our real-world view of what a stack is. Certainly, because the datatype declaration is so powerful in coercing types, we have to be careful not to overdo it and create nonsensical types like our original infinitely regressing stack, but we shall find that the advantages of creating **dynamic types** far outweigh the disadvantages of bending the rules slightly.

Now we have invented the dynamic type *'a stack*, we can model a pile of crockery, for instance. Let's restrict the values of type crockery to be plates and saucers; we have:

```
datatype crockery = PLATE | SAUCER;

datatype crockery
constructor PLATE : crockery
constructor SAUCER : crockery
```

Now, starting with the empty stack

```
EMPTY_STACK;

val it = EMPTY_STACK : 'a stack
```

we can add a plate:

```
PLATE PUSH it;

val it = PUSH(PLATE, EMPTY_STACK) : crockery stack
```

and a saucer:

```
SAUCER PUSH it;

val it = PUSH(SAUCER, PUSH(PLATE, EMPTY_STACK))
: crockery stack
```

We now have a stack (*it*) which we can use in expressions like any other value, which has inside it a SAUCER and a PLATE waiting to be released by a *pop* operation.

The *pop* Operation

The *pop* operation is the inverse of *push* (Fig. 7.3). We start with a *'a stack* and end up with an item of type *'a* and a *'a stack*. We have

$$pop : 'a\ stack \rightarrow 'a * 'a\ stack$$

pop is not a constructor function — quite the reverse — so we can use an ordinary ML function to model it:

```
fun pop (a PUSH A) = (a,A);

ML WARNING - Clauses of function binding are
non-exhaustive
```

```
INVOLVING: fun pop
val pop = fn : 'a stack -> 'a * 'a stack
```

Fig. 7.3 Popping an item off a stack

Our PUSH constructor is of arity 2, and we enclose it in brackets to avoid confusing ML. But ML is still not happy, as we have not dealt with all the possible values of the type *'a stack*. What happens when we try to *pop* an empty stack?

Well, it is meaningless to *pop* from an empty stack, but because we have coerced all sizes of stack to be of the same type, we have to find a way of dealing with the situation. One solution is to raise an exception in this case:

```
exception Pop
fun pop (a PUSH A)   = (a,A)
|   pop EMPTY_STACK = raise Pop
;
exception Pop
val pop = fn : 'a stack -> 'a * 'a stack
```

Note that we have put the empty stack case last, as it is less likely to be met in practice. Now we can retrieve our crockery:

```
it;

val it = PUSH(SAUCER, PUSH(PLATE, EMPTY_STACK))
: crockery stack

val (item1,it) = pop it;

val item1 = SAUCER : crockery
val it = PUSH(PLATE, EMPTY_STACK) : crockery stack

val (item2,it) = pop it;

val item2 = PLATE : crockery
val it = EMPTY_STACK : crockery stack
```

```
val (item3,it) = pop it;

ML EXCEPTION: Pop
```

It would be very handy if we could find out whether a stack was empty before popping it. Because the components of types can be matched inside a function, this is no problem:

```
fun is_empty_stack EMPTY_STACK = true
|   is_empty_stack _           = false
;

val is_empty_stack = fn : 'a stack -> bool
```

In this function we have put the empty stack case first, because the other case is a wild card which will match anything, even the empty stack!

```
is_empty_stack it;

val it = true : bool
```

7.6 PROOF OF PROPERTIES OF THE TYPE *stack*

We have been rather informal in our treatment of the type stack, and it's now time to prove some properties of the type. We shall use a method called **algebraic specification** to describe the properties we wish to prove. For a stack, the specification is very simple: we have to show that the last item pushed is the first item to be popped — a property known as **last in first out** or LIFO. This is a necessary property of any type that we want to call a stack. Particular variants of stacks may have additional properties, but they must all be LIFO to warrant the name *stack*.

The property can be formulated as follows: given an arbitrary stack A, and an item a , if we push a onto the stack, we have a situation where popping the stack will give us a and the stack A. In formula:

$$pop\ (a\ push\ A) \quad = \quad (a,A)$$

This is the algebraic specification of a stack. To prove that our implementation satisfies this specification we compare it with our definition of *pop*:

Proof

```
exception Pop
fun pop (a PUSH A)          = (a,A)
|   pop EMPTY_STACK         = raise Pop
;
```

We can see that the first line of the definition is of the same form as the specification. In this case the program *is* the proof.
QED

The *satisfies* Relationship

There is, of course, some extra behaviour in our implementation which is not specified by the algebraic specification. No matter! The implementation **satisfies** the specification if it behaves as the specification says it should. We can see that this relation of satisfaction between a specification and an implementation is not necessarily one-to-one. We can imagine many different implementations for one specification; for example, we could have an implementation in which attempting to pop an empty stack causes the entire system to halt. So we have a many-to-one mapping between implementations and specifications, and we can define a function (sometimes called a **retrieve function**) which takes an implementation as argument and returns the equivalent specification. By contrast, the relationship between specifications and implementations is, in general, one-to-many. We cannot take a specification, apply a function to it, and obtain an implementation. Implementing a specification is not an algorithmic process which could be carried out by an automaton, it requires original thought.

7.7 QUEUES

A queue is another familiar dynamic type from daily life (Fig. 7.4). Like a stack, it can be empty, and it can extend to any length. Unlike a stack, the first item in the queue is the first item out of it. This property is known as FIFO or **first in first out**.

This time let us define the queue formally first and try to derive the ML code from it. Then a proof will be unnecessary. We shall talk about **putting** an item onto a queue, and **getting** an item from a queue. We can make *put* an infix function as before, this time of type $'a\ queue * 'a \rightarrow 'a\ queue$. As we are adding to the right-hand side of the queue, *put* will be left associative. A typical queue will look like

empty_queue *put* Alice *put* Brian *put* Clarissa

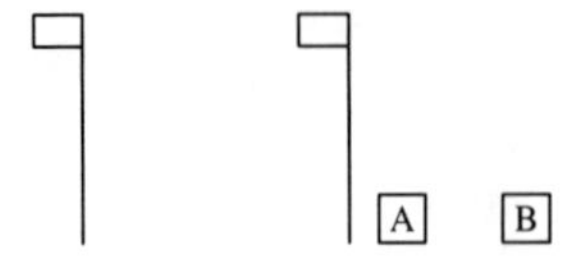

Fig. 7.4 An empty queue/a queue of two items

get will have type *'a queue* → *'a * 'a queue*. A *get* will convert the above queue into

(Alice, empty_queue *put* Brian *put* Clarissa)

How it does this is not immediately clear. In fact we need a recursive definition. *get* (empty_queue) will be undefined, like *pop* (empty_stack), so the base case occurs when there is just one item on the queue:

get (empty_queue *put* a) = (a,empty_queue)

For the inductive case we use the method of differences: how does the result of a *get* change as the length of the queue is increased by 1? Figure 7.5 may be helpful.

The item taken off is exactly the same in the two cases. The only change is that the newly added item remains on the end of the queue when the first item is taken off. In formula

if $\quad get\ A \;=\; (a,B)$

then $get(A\ put\ d) \;=\; (a,B\ put\ d)$

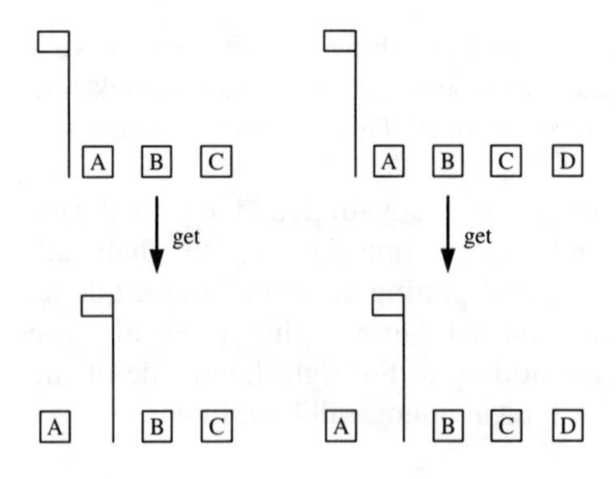

Fig. 7.5 The effect of *get* on queues of increasing length

In effect we have defined the result of *get*(*A put d*) in terms of *get A*, so we have a valid recursive definition for a queue. We now implement this specification in ML. First, we need a datatype for the type *'a queue*: We have a nullary constructor EMPTY_QUEUE, and the binary constructor PUT, which is left associative and of precedence 5:

```
infix 5 PUT
datatype 'a queue = EMPTY_QUEUE
                  | op PUT of 'a queue * 'a
;
infix 5 PUT
datatype 'a queue
constructor EMPTY_QUEUE : 'a queue
constructor op PUT : 'a queue * 'a -> 'a queue
```

This declaration is very similar to the one for *'a stack*, apart from the items being added at the opposite end.

```
datatype person = ALICE | BRIAN;

datatype person
constructor ALICE : person
constructor BRIAN : person

EMPTY_QUEUE;

val it = EMPTY_QUEUE : 'a queue

it PUT ALICE;

val it = PUT(EMPTY_QUEUE, ALICE) : person queue

it PUT BRIAN;

val it = PUT(PUT(EMPTY_QUEUE, ALICE), BRIAN) :
person queue
```

Now we need to define *get*. The formal specification

get (empty_queue *put a*) = (*a*,empty_queue)

if *get A* = (*a*,*B*)
then *get*(*A put d*) = (*a*,*B put d*)

can be converted to ML with the help of the *let* expression, which is used to analyse the result of *get A*. Once again, we use an exception to deal with the undefined case of attempting to remove something from an empty queue:

```
exception Get
fun get EMPTY_QUEUE               = raise Get
|   get (EMPTY_QUEUE PUT a) = (a,EMPTY_QUEUE)
|   get (A PUT d)     =
  let
    val (a,B) = get A
  in
    (a,B PUT d)
  end
;
exception Get
val get = fn : 'a queue -> 'a * 'a queue
```

Now we can unqueue Alice and Brian:

```
val (item1,it) = get it;

val item1 = ALICE : person
val it = PUT(EMPTY_QUEUE, BRIAN) : person queue

val (item2,it) = get it;

val item2 = BRIAN : person
val it = EMPTY_QUEUE : person queue

val (item3,it) = get it;

ML EXCEPTION: Get
```

Alice and Brian come off the queue in the order they went on, in marked contrast to the stack of plates and saucers. Once again we can avoid the exception by inventing a function to detect the null case:

```
fun is_empty_queue EMPTY_QUEUE = true
|   is_empty_queue _           = false
;

val is_empty_queue = fn : 'a queue -> bool

is_empty_queue it;

val it = true : bool
```

7.8 INTEGRATING THE TYPES *stack* AND *queue*

Our types *'a stack* and *'a queue* have a lot in common. If we wanted to extend the operations we perform on them, and for example equate two stacks or two queues, or find the length of a stack or queue, we would find even more common ground. There is a reason for this. The type *'a stack* with its operation *push*, and the type *'a queue* with its operation *put*, are **isomorphic**, that is, there is a bijective mapping between values of the two types:

empty_stack	$\leftrightarrow$	empty_queue
a push empty_stack	$\leftrightarrow$	empty_queue *put a*
b push a push empty_stack	$\leftrightarrow$	empty_queue *put a put b*

and so on. The isomorphism extends to some of the other operations too: for example, *is_empty_queue* and *is_empty_stack*, and the length and equality operations (but not to *get* and *pop*). So, under our motto of *review and improve*, is there no way in which we can combine the two types?

7.9 LISTS

We have met the *list* type briefly in Chapter 4. We saw examples of constant lists, such as [1,2,3] and [], and invented a (partial) function, *reverse*, which reversed a list of three members:

```
fun reverse [a1,a2,a3] = [a3,a2,a1];

ML WARNING - Clauses of function binding are
non-exhaustive

val reverse = fn : 'a list -> 'a list
```

list is a polymorphic type, like *stack* and *queue*, and furthermore, it is isomorphic with them:

empty_stack	$\leftrightarrow$	[]	$\leftrightarrow$	empty_queue
a push empty_stack	$\leftrightarrow$	[*a*]	$\leftrightarrow$	empty_queue *put a*
		....		

so it seems a good candidate for combining the two types.

I shall first describe the data type *list*, as it is implemented in ML, and then we shall attempt to implement *stack* and *queue* using it.

The *list* type is constructed like the *stack* type. There is a standard ML constructor called *nil* which represents the empty list, and a standard infixed constructor called :: (usually pronounced 'cons' or 'prefix') which prefixes a value to a list:

```
infixr 5 :: ;
datatype 'a list = nil
                 | op :: of 'a * 'a list
;
infixr 5 ::
datatype 'a list
constructor nil : 'a list
constructor op :: : 'a * 'a list -> 'a list
```

This type is pre-declared in every ML implementation. It allows lists to be built up just like stacks:

```
nil;

val it = [] : 'a list

3 :: it;

val it = [3] : int list

2 :: it;

val it = [2, 3] : int list

1 :: it;

val it = [1, 2, 3] : int list
```

ML chooses to output 1::2::3::*nil* in the more compact form [1,2,3], but the two notations mean exactly the same thing.

Implementing *stack* in Terms of *list*

Now it is easy to implement the type *stack* in terms of the type *list*. The EMPTY_STACK constructor becomes the value *empty_stack*, defined in terms of the null list:

```
val empty_stack = [];

val empty_stack = [] : 'a list
```

Note that *empty_stack* is no longer a constructor itself, so it is not available for pattern-matching. The constructor PUSH becomes an ordinary function, *push*, defined in terms of the :: constructor:

```
infixr 5 push
fun a push A = a::A;
infixr 5 push
val op push = fn : 'a * 'a list -> 'a list
```

and the *pop* operation is defined by pattern-matching on null and non-null lists:

```
exception Pop
fun pop (a::A) = (a,A)
|   pop [] = raise Pop
;

exception Pop
val pop = fn : 'a list -> 'a * 'a list
```

Proof of the Validity of the Implementation of *stack*

I am making unsubstantiated claims here for my implementation, so let's prove that the above code does satisfy our specification for *stack*:

$$pop\ (a\ push\ A) \quad = \quad (a,A)$$

We have

	$pop(a\ push\ A)$	
=	$pop(a::A)$	definition of *push*
=	(a,A)	definition of *pop*
QED		

Implementation of *queue* in Terms of *list*

Again, this implementation is very straightforward. The empty queue is implemented as the null list:

```
val empty_queue = [];

val empty_queue = [] : 'a list
```

The *put* operation is implemented in terms of the :: constructor:

```
infix 5 put
fun A put a = a::A;

infix 5 put
val op put = fn : 'a list * 'a -> 'a list
```

Only the *get* operation requires some mental effort. As we are putting items onto the left-hand end of the queue, we must get them from the right-hand end. This suggests a recursive definition of getting, very similar to our original one:

```
exception Get
fun get []       = raise Get
|   get [a]      = (a,[])
|   get (a::A)  =
  let
    val (b,B) = get A
  in
    (b,a::B)
  end
;

exception Get
val get = fn : 'a list -> 'a * 'a list
```

Proof of the Validity of the *list* Implementation of *queue*

Once again, we show that the implementation satisfies our specification:

get (empty_queue *put a*) = (*a*,empty_queue)
if *get A* = (*a*,*B*)
then *get*(*A put d*) = (*a*,*B put d*)

Proof for queue of one item

	get (empty_queue *put a*)	
=	*get*([] *put* a)	definition of empty_queue
=	*get*([*a*])	definition of *put*
=	(*a*,[])	definition of *get*
=	(*a*,empty_queue)	definition of empty_queue

Proof for queue of many items

assume	$get\ A = (a,B)$	(1)
then	$get(A\ put\ d)$	
=	$get(d::A)$	definition of *put*
=	$(a,d::B)$ where $(a,B) = get\ A$	definition of *get* and assumption (1)
=	$(a,B\ put\ d)$	definition of *put*
QED		

Additional Operations on Type *list*

Having implemented both *stack* and *queue* successfully in terms of type *list*, we are free to implement any other useful operations we require on the type *list* alone. For example, the *is_empty* operation:

```
fun is_empty [] = true
|   is_empty _  = false
;

val is_empty = fn : 'a list -> bool
```

which implements both *is_empty_list* and *is_empty_queue*. The *length* function is easily designed by the method of differences — the length of a list increases by one every time we prefix a new item, and the length of the null list is zero:

```
fun length (a::A) = 1 + length A
|   length []     = 0
;

val length = fn : 'a list -> int
```

And so on . . . there is a wealth of operations we may wish to perform on the type *list*, and we shall explore them in more detail in the next chapter.

7.10 PROBLEMS OF COMBINING TYPES

Although combining types in this way is very convenient, and has great potential for reducing redundancy in programs, there are a few caveats. In the above implementation, there is nothing to stop us getting and putting from a value that is nominally a stack, or popping an item from a value that we have previously treated as a queue. In other words, we have sacrificed some type security for economy and efficiency.

This is usually not serious for a small program written by one person, but in large projects written by a team of programmers incompatibilities can arise which can destroy the integrity of the system. It can be very hard to spot the one line of code in 10 000 where someone has bent the rules slightly, yet this sort of error is distressingly frequent in present-day software. (With amazing pertinence, the word-processing package I am using to type this book came to an unexpected halt at precisely this point. Fortunately it saved the text file first!)

Because of the disastrous effect that even a small breach of type security can have, and the high cost of detecting and correcting such errors, a movement has grown up in the past decade or so which prizes type security above all else. For this school of thought, doing what we have done in the last few sections is the deadliest sin. There are three points to be made in reply to this criticism.

First, it can happen that the real-world phenomenon which we are modelling using type A turns out to be of type B; in other words, we may actually want to pop something from a queue occasionally, quite legitimately. In a totally type-strict program, this would mean converting the type *queue* to the type *double ended queue* or somesuch, and re-implementing all its operations. This is also a high-cost operation.

Second, everything we write in ML is being proved against a formal specification. This has two consequences: it should be impossible to implement anything that is incompatible with the specification; and modifications to the system, when they occur, should be carried out as changes to the specification, which then feed through into the implementation. As long as the specification is type-secure, the implementation will also be type-secure, even if types are combined.

Finally, type-strictness is limited by the types inherent in the implementation language. For example, there is no type *natural_number* in ML. A system involving natural numbers will be *specified* in terms of natural numbers, but *implemented* in terms of the ML type *int* (in other words, the type *natural_number* combined with the type *negative_number*). This presents no problem if the implementation is proved to satisfy the specification.

Of course, at the present time we are a long way from the situation epitomized in the last two paragraphs. Fixes to software are often carried out hastily, without reference to any specification and with inadequate testing, let alone proof. But a more rigorous mode of development has to be admitted as a possibility. I hope this book goes some way to showing that it is a practical possibility, at least for functional languages.

Type-secure Implementation of *stack* in Terms of *list*

To see what is at stake here, let us implement the type *stack* in terms of *list*, but in a type-secure way. We will first of all need a conversion function to construct the type *'a stack* from the type *'a list*:

```
datatype 'a stack = STACK of 'a list;

datatype 'a stack
constructor STACK : 'a list -> 'a stack
```

Then we need to implement *empty_stack* and *push* in a type-secure way, using STACK to do the type conversions:

```
val empty_stack = STACK([]);

val empty_stack = STACK([]) : 'a stack

infixr 5 push
fun a push (STACK A) = STACK(a::A);

infixr 5 push
val op push = fn : 'a * 'a stack -> 'a stack
```

Now *pop* must be implemented in the same type-secure way:

```
exception Pop
fun pop (STACK (a ::A)) = (a,STACK A)
|   pop (STACK [])      = raise Pop
;

exception Pop
val pop = fn : 'a stack -> 'a * 'a stack
```

and all the other functions we would like to use must also be given type-secure implementations:

```
fun is_empty_stack (STACK A) = is_empty A;

val is_empty_stack = fn : 'a stack -> bool

fun length_stack (STACK A) = length A;

val length_stack = fn : 'a stack -> int
```

and so on. Now, if we do all this again for the type *queue*, we will not be able to say

```
val A = "x" push empty_stack;

val A = STACK(["x"]) : string stack

val (a,A') = get(A);

ML TYPE ERROR - Type unification failure
WANTED : 'ty1 queue
FOUND  : string stack
```

There is no doubt that the strict type-checking of this approach can detect many errors of implementation. Also, because this type-checking is done before the translation of the ML program into machine code, there need be no resulting inefficiency in the running program. And yet, and yet . . . the duplication of code, though trivial, is irksome. You will have to decide, in each situation, whether the full panoply of strict type-checking is necessary. ML provides the facilities for you to make that choice.

7.11 SYNTAX USED IN THIS CHAPTER

Expressions

atexp	::=	`let` *dec* `in` *exp* `end`	local declaration

Declarations

dec	::=	`datatype` *tyvar tycon* = *conbind*	
		dec_1 dec_2	sequential declaration
		`infix` ⟨*d*⟩ *id*	left-associative infix
		`infixr` ⟨*d*⟩ *id*	right-associative infix
conbind	::=	⟨op⟩ *con* ⟨ `of` *ty*⟩ ⟨	*conbind*⟩	

Patterns

pat	::=	*atpat*	atomic
		⟨op⟩ *con atpat*	value construction
		pat_1 *con* pat_2	infixed value construction
		pat : *ty*	typed

7.12 CHAPTER SUMMARY

As well as constant constructors, ML has constructor functions. These can be defined using a *datatype* declaration to create new types from old, and to coerce several types into one. Because constructor functions can be used in patterns on the left-hand side of declarations, it is easy to deconstruct the resulting types using pattern-matching. To make their special status clear, we use CAPITAL LETTERS for constructor function names.

Constructor functions come into their own when we use dynamic types like *stack*, *queue* and *list*. These types are polymorphic and can have components of any type, but all the components of a given instance of a dynamic type must be of the same type, so we can have an *integer queue*, *string list* and so on.

The *let* expression can be used to avoid multiple evaluation of the same expression, or to analyse complex expressions.

As many dynamic types are similar or even isomorphic, we have to make a trade-off between type security and compactness when we implement them. A formal specification can help in this dilemma, as the type strictness can be confined to the specification.

We use an algebraic method of formal specification in this book. One specification can give rise to many implementations; we say the implementation *satisfies* the specification.

EXERCISES

1. Define a datatype *marks* with an associated constructor function MARKS of type *real→marks* and a datatype *pounds* with an associated constructor function POUNDS of type *real→pounds*.
2. Write a conversion function *pounds_marks* of type *pounds→marks*, assuming there are 2.40 marks to the pound.
3. In a spirit of European unity, construct a datatype *currency* which can either be francs, marks, lire, pesetas or pounds (apologies to the rest

of Europe, it would make the exercise too long). The argument types of the constructor functions should all be *real*.

4. Write a function *ecuise* which will convert an argument of type *currency* to a result of type *ecu*. Assume the following conversion factors:

1 ecu	=	13.9 francs
1 ecu	=	4.21 marks
1 ecu	=	3500 lire
1 ecu	=	256 pesetas
1 ecu	=	1.93 pounds

5. Define a constructor function PAIR which constructs a pair of values out of two values of the same type.
6. Define a constructor TRIPLE which constructs a triple of values out of three values of the same type.
7. Write a function *first_of_pair* which returns the first item of a pair defined by the constructor PAIR.
8. Write a function *last_of_triple* which returns the last item of a triple defined by the constructor TRIPLE.
9. You have been given the job of designing the software system for collecting passenger vehicle tolls on the Channel Tunnel. The system has to cope with three kinds of passenger vehicle: bikes, motorbikes and cars. Bikes pay a fixed fee of £10. Motorbikes pay a fee of £10 plus 60p for each 100 cubic centimetres of engine capacity. Cars pay a fee of £20 (£15 for three-wheeled vehicles) plus £1.50 pro rata for each metre of length in excess of 3 metres.
   (a) Write datatype declarations to specify the following types:
       - Cubic centimetres (*cc*)
       - Length in metres (*metres*)
       - Number of wheels (*wheels*)
       - Fee in pounds (*pounds*)
       - Passenger vehicles (*vehicle*)

   (b) Write a function *toll* which, given an argument of type *vehicle*, returns the appropriate fee in *pounds*.
   (c) On the other side of the tunnel, of course, the toll will be collected in francs. Assuming the same rates apply, and there are 8.23 francs to the pound, enhance your system so that given a *vehicle* it can give the fee in pounds and francs.
10. Newton's method for obtaining the square root r of a real number x makes an initial guess r_1 for the root, and then uses the formula

$$r_{i+1} = 0.5(x/r_i + r_i)$$

to obtain the next approximation to the root. It can be shown that this formula always **converges** or produces better and better approximations to the root, for *any* (positive) starting value. Using this idea, invent a function *my_sqrt* which will give the square root of a real number accurate to three decimal places. Check your function by using easy values for x such as 25.0, 36.0, etc.

11. Given a type *traffic light* defined as follows:

```
datatype light =
  RED | RED_AMBER | GREEN | AMBER;
```

a type sensor defined :

```
datatype sensor = TRUE | FALSE;
```

and a switching function defined:

```
val switch =
  fn RED        => RED_AMBER
   | RED_AMBER  => GREEN
   | GREEN      => AMBER
   | AMBER      => RED
;
```

write a function *change* that takes two sensor values and two light values and returns two updated sensor values and two updated light values. The function is supposed to model a crossroads with a sensor and light on each road (actually, of course, there are four sensors and four lights, but life is too short to worry about this). The function will be easier to write if you use a *let* expression.

12. Given a stack defined as follows:

```
datatype 'a stack = NULL
                  | PUSH of 'a * 'a stack
;
```

Define the following functions:

(a) *stack_of* which creates a stack of n identical items of value a. Think carefully about the type of function *stack_of*.
(b) *last_item* which gives the last item in a stack.
(c) *all_but_last* which returns a stack minus its last item.
(d) *invert_stack*, which inverts a stack, so the last item becomes the first and so on. (The previous two functions may be useful.)
(e) *swap* which swaps the top two items of a stack.

(This function is partial, so you'll have to work out the conditions under which an exception is raised. The use of pattern-matching and wild cards is recommended.)

(f) *take* which returns the first *n* items on a stack.
(First of all think about the type of this function — there are two arguments and you may want to make them curried. The Method of Differences is required here. Think carefully about the base case — or is it cases? How will your function deal with negative *n*?)

13. We could have defined *queue* in terms of *list* in another way — adding items on the right-hand side and removing them from the left-hand side. Redefine *queue* in this way, and prove that it still satisfies the specification.
14. Using only the standard ML functions:

```
infixr 5 ::
::   : 'a * 'a list -> 'a list
```

which prefixes an item to a list,

```
infixr 5 @
@    : 'a list * 'a list -> 'a list
```

which concatenates two lists into one, and

```
rev : 'a list -> 'a list
```

which reverses a list, design the following functions:

(a) *first* which returns the first item of a list, and *rest* which returns all but the first item of a list. (Think about the types of these functions first of all, and then how they would deal with a null list and a non-null list.)
(b) *last* and *front*, which give the last item and all but the last item of a list, respectively.
(c) *postfix* which appends a value to a list. (First write down its type, then see if you can express it in terms of the standard functions.)

CHAPTER

EIGHT

LIST PROCESSING

8.1 INTRODUCTION

The first functional language, Lisp, was dedicated to list processing, and lists still loom large in functional programming today. There is a bewildering array of operations that can be performed on lists, so in this chapter I have attempted to classify them and show the relationships between them.

A few notational conventions to begin with. A list will normally be denoted by a capital letter which corresponds to the lower-case letter used for its constituent type, so a list of integers would be N, a list of reals X, a list of strings S, and a list of Booleans P. General lists of type *'a*, *'b*, etc. are given the names A, B, etc. The individual items of a list are denoted by subscripted lower-case letters, so list P would consist of $[p_1, p_2, \ldots, p_n]$. An arbitrary item of list A will be denoted by a, and so on. These are all the conventions we shall need for our study of lists.

8.2 FOLDING

Most operations on lists are going to be repetitive in nature. We have seen one of them already:

```
fun length (a::A)  = 1 + length A
|   length []      = 0
;

val length = fn : 'a list -> int
```

Here is a function that finds the product of all the items in an integer list:

```
fun product (n::N)  = n * product N
|   product [] = 1
;

val product = fn : int list -> int
```

Here is another function, called *map*, which maps each item of a *'a list* to the corresponding item of a new *'b list*, according to some function $f : 'a \to 'b$:

```
fun map f (a::A)  = f a :: map f A
|   map f []      = []
;

val map = fn : ('a -> 'b) -> 'a list -> 'b list
```

so *map* (*fn n* => 2 * *n*) is the function that doubles all the items in an integer list.

Finally, here is a function to filter out all the values of a list for which a certain condition, represented by some function $f : 'a \to bool$, is true:

```
fun filter f (a::A) = if f a then a :: filter f A
                                 else filter f A
|  filter f []       = []
;

val filter = fn : ('a -> bool) -> 'a list -> 'a list
```

so *filter* (*fn n* => *n mod* 2 = 0) is the function that filters the even items out of an integer list, and

filter (*fn n* => *n mod* 2 = 0) [1,2,3] = [2]

In all these examples of list processing functions we see the familiar inductive case and base case, and it is clear that an inductive proof is going to be very similar for all four functions. Under these circumstances the thought naturally occurs: can we produce a generalized function for induction on a list as we did for induction on the natural numbers in Chapter 6?

In fact there is a very strong analogy between the two kinds of induction. Just as in Chapter 6 we ended up with two functions for induction on the natural numbers, *compose* and *iter*, which were almost equivalent, for induction on a list we also find two functions, *foldr* and *foldl*, which again are almost equivalent. Once again, efficiency considerations will play a part in our decision to use one or other of the folding functions.

Let's first decide exactly what we mean by induction on a list. For induction on the natural numbers, we used the idea of composing a function n times, f^n, where n is a natural number. We then applied f^n to some argument a to obtain our required result, $f^n\ a$. Is there an equivalent expression for induction on a list?

Well, induction on the null list is going to be analogous to f^0. In the natural number case, the result was just the initial argument, but we can see that this is not always going to be the case for lists:

length [] = 0
product [] = 1
map (*fn n* => 2 **n*) [] = []
filter (*fn n* => *n mod* 2 = 0) [] = []

Also, the non-null case does not consist of just repeatedly applying a function to an argument, but in some way applying a function to every item of the list. The key to understanding the idea of folding is to imagine an infix operation (conventionally called ⊕) between each item of the list:

$$\ldots \oplus a_j \oplus a_{j+1} \oplus a_{j+2} \oplus \ldots$$

To deal with the null list case, we have to put the identity element for the operation on one end of the list, giving:

$$i \oplus a_1 \oplus a_2 \oplus \ldots \quad \ldots \oplus a_{n\text{-}1} \oplus a_n$$

So, for example, for adding all entries in the list, we would have:

$$0 + a_1 + a_2 + \ldots \quad \ldots + a_{n\text{-}1} + a_n$$

and for multiplying all entries in the list, we would have:

$$1 \times a_1 \times a_2 \times \ldots \quad \ldots \times a_{n\text{-}1} \times a_n$$

These are general expressions which give the required result for any length of list, including the null list.

As we want to compute the results of these expressions, we have to specify the order in which the operations are carried out. By convention, we start from the end with the identity element, so the above expressions would be evaluated in the order

$$((\ldots((i \oplus a_1) \oplus a_2) \oplus \ldots \quad \ldots \oplus a_{n\text{-}1}) \oplus a_n)$$

This is **folding from the left**. There is also the possibility of putting the identity element at the right-hand end, and evaluating from the right:

$$(a_1 \oplus (a_2 \oplus \ldots \quad \ldots \oplus (a_{n\text{-}1} \oplus (a_n \oplus i))\ldots))$$

This is **folding from the right**.

Now we design the two folding functions, which we shall call *foldl* for folding from the left and *foldr* for folding from the right. First let's determine their types. Each will take a list as argument, of course, and this list will have an arbitrary type as item, so it will be a *'a list*. We shall also need the infix operation, $\oplus$, and the identity element, i, as arguments. Our first thought is that $\oplus$ will have the type *'a * 'a -> 'a*, and i the type *'a*, and so they will for **commutative** operations like + and $\times$, where the arguments can be swapped around without changing the result, i.e. $a \oplus b = b \oplus a$. For this kind of operator, we could say

$$foldl \oplus i\ A = a \qquad \text{and} \qquad foldr \oplus i\ A = a$$

and *foldl* and *foldr* would each have the type

$$('a * 'a \rightarrow 'a) \rightarrow 'a \rightarrow 'a\ list \rightarrow 'a$$

But the folding process is more general than this. Suppose we are folding from the left:

$$((\ldots((i \oplus a_1) \oplus a_2) \oplus \ldots \quad \ldots \oplus a_{n\text{-}1}) \oplus a_n)$$

and suppose i and a_1 have two different types, *'a* and *'b*. The result of applying the $\oplus$ operator $(i \oplus a_1)$ is going to be the left-hand argument of the next $\oplus$ operator, so it must have the type *'a*. This gives us the general type $'a * 'b \rightarrow 'a$ for the $\oplus$ operator when we fold from the left.

Similarly, when we fold from the right:

$$(a_1 \oplus (a_2 \oplus \ldots \quad \ldots \oplus (a_{n\text{-}1} \oplus (a_n \oplus i))\ldots))$$

we start by applying the $\oplus$ operator to a_n and i. If we imagine these to be of different types, $'a$ and $'b$, then the result must be of type $'b$ to be a valid right argument to the next application of $\oplus$. So we end up with $\oplus$ having the type $'a * 'b \rightarrow 'b$ when we fold from the right.

The type of $\oplus$ determines the type of i and A. For *foldl*, i must have type $'a$, and A must be of type $'b$ *list*. The result will be of type $'a$. We have

$$foldl \ \oplus \ i \ B \ = \ a$$
$$foldl : ('a * 'b \rightarrow 'a) \rightarrow 'a \rightarrow 'b \ list \rightarrow 'a$$

Similarly, for *foldr*, i must be of type $'b$ and A must be of type $'a$ *list*, and the result will be of type $'b$:

$$foldr \ \oplus \ i \ A \ = \ b$$
$$foldr : ('a * 'b \rightarrow 'b) \rightarrow 'b \rightarrow 'a \ list \rightarrow 'b$$

Note that in both cases, the result can be of a completely different type from the list, and may not be a list at all. For instance, we could dream up an operator *mult_real* : *real* * *int* $\rightarrow$ *real*, which takes a real and an integer and multiplies them together to give a real result. We could then fold this operator over an integer list:

$$foldl \ mult\_real \ 2.0 \ [1,2,3] \ = \ 12.0$$

But what are these strange operators that take different left and right arguments? They are clearly not commutative. They are also not **associative**, i.e.

$$(a \oplus b) \oplus c \ \neq \ a \oplus (b \oplus c)$$

In fact we say that an operator of type *'a * 'b -> 'a* is **left associative**, or associates to the left, and an operator of type *'a * 'b -> 'b* is **right associative**, or associates to the right. For left-associative operators we must fold from the left, and for right-associative operators we must fold from the right. For fully associative operators (like $\times$) we can fold either way, and we have the identity

$$foldl \ \oplus \ i \ A \ = \ foldr \ \oplus \ i \ A$$

where $\oplus$ is a fully associative operator.

Folding from the Left

We want to invent a function

$$foldl : ('a *'b \rightarrow 'a) \rightarrow 'a \rightarrow 'b\ list \rightarrow 'a$$

such that the expression *foldl* $\oplus$ *i A* will evaluate to

$$((\ldots((i \oplus a_1) \oplus a_2) \oplus \ldots \quad \ldots \oplus a_{n-1}) \oplus a_n)$$

To comply with the syntax rules of ML, the infixed operator $\oplus$ has to be converted to a prefixed function *f*, giving us the slightly more opaque formulation

$$foldl\ f\ i\ A \;=\; \ldots f(f(f(i,a_1),a_2),a_3)\ldots$$

To design the function we use the Method of Differences. The base case is simple; *A* is a null list and the result is the identity element:

```
foldl f i [] = i
```

Now we must find a recurrence relation between the result of folding list *A* and the result of folding list (*a*::*A*). From the definition of folding from the left we can see that folding list (*a*::*A*) with identity element *i* will give the same result as folding list *A* with identity element *f*(*i*,*a*), so we have

```
foldl f i (a::A) = foldl f (f(i,a)) A
```

giving the complete ML definition:

```
fun foldl f i (a::A) = foldl f (f(i,a)) A
|   foldl f i []     = i
;

val foldl = fn : ('a * 'b -> 'a) -> 'a -> 'b list -> 'a
```

This is our first formulation of the idea of repetitive operations on a list. Let's check it out. We can use summing a list as our test example. The $\oplus$ operation will be + and the *i* element will be 0:

```
foldl + 0 [1,2,3,4];

ML TYPE ERROR - Type unification failure
```

```
WANTED    :  int list -> 'ty1
FOUND     :  int
```

What has happened here? The answer lies in ML's type inferencing process. Unfortunately, because + is an infix operation, ML has taken *foldl* + 0 [1,2,3,4] to be an infixed expression. It assumes 0 [1,2,3,4] is a function application, and recognizes [1,2,3,4] as an integer list, so it infers that 0 should have type *int list -> 'ty1*, where *'ty1* is some as yet unknown type. It is then terribly disappointed to find that 0 actually has type *int*.

We must make it clear to ML that + in this context is just an ordinary function of type $'a * 'a \to 'a$, a prefixed function in fact. We know how to do this — prefix it with *op*.

```
foldl op+ 0 [1,2,3,4];

val it = 10 : int
```

Success! Now let's try a less obvious example, like successive division of an integer by the items of a list. This time *i* will be the integer to be divided, and ⊕ will be the *div* operator:

```
foldl op div 144 [1,2,3,4];

val it = 6 : int
```

It appears to work. Note that when *op* prefixes a non-alphabetic function name such as +, the intervening space can be omitted to give a single symbol such as *op+*. Even when *op* prefixes an alphabetic function name, the two words are treated as a single symbol by the ML interpreter, so *op div* is treated as a single function name. This means that brackets are never necessary, even when, as in the example above, *op div* is the argument of a curried function.

Folding from the Right

This time we want a function *foldr* such that

$$foldr \oplus i\ A = (a_1 \oplus (a_2 \oplus \ldots \quad \ldots \oplus (a_{n-1} \oplus (a_n \oplus i))\ldots))$$

or in ML terms

$$foldr\ f\ i\ A = f(a_1, \ldots f(a_{n-2}, f(a_{n-1}, f(a_n, i)))\ldots)$$

For the base case we have as before

```
foldr f i [] = i
```

but the recurrence relation between A and $(a::A)$ is slightly different. This time the result of folding $(a::A)$ with identity element i will be the same as the result of folding A with identity element i and then applying f to a and the result:

```
foldr f i (a::A) = f(a,foldr f i A)
```

giving us the full definition:

```
fun foldr f i (a::A)  = f(a,foldr f i A)
|   foldr f i []      = i
;

val foldr = fn : ('a * 'b -> 'b) -> 'b -> 'a list -> 'b
```

— our second formulation of the idea of repetition on a list. We quickly check it for fully associative operators:

```
foldr op+ 0 [1,2,3,4];

val it = 10 : int

foldr op* 1 [1,2,3,4];

val it = 24 : int
```

Once again, it seems to work satisfactorily.

Which Way to Fold?

From the form of the definitions for *foldl* and *foldr*, we can see that *foldl* is tail recursive and *foldr* is not, so *foldl* will always give a more efficient implementation in cases where a choice is possible. For example, we could find the length of a list without using explicit recursion by using *foldl*. We have already seen the explicit version of *length*:

```
fun length (a::A)  = 1 + length A
|   length []      = 0
;
```

```
val length = fn : 'a list -> int
```

All that is required for the implicitly recursive version is the form of the ⊕ operator and the *i* element. The *i* element is easy; it is zero. The ⊕ operator is going to take an integer and a list item of type *'a*, and give an integer as result. What we really want to do is take a list

$$[a_1,a_2, \ldots a_{n-1},a_n]$$

and convert it into the sum

$$0 + 1 + 1 + \ldots +1 + 1$$

For folding from the left, the ⊕ operator will have type *int * 'a* → *int*, and will simply add one to its left argument, ignoring the right argument completely:

```
val f = fn (n,a) => n+1;

val f = fn : int * 'a -> int
```

so we can define

```
val length = foldl f 0;

val length = fn : 'a list -> int
```

Now let's evaluate a simple expression using the *foldl* version of *length*:

```
length ["a","b","c"]
=     foldl f 0 ["a","b","c"]
=     foldl f (f(0,"a")) ["b","c"]
=     foldl f 1 ["b","c"]
=     foldl f (f(1,"b")) ["c"]
=     foldl f 2 ["c"]
=     foldl f (f(2,"c")) []
=     foldl f 3 []
=     3
```

and once again using the explicit version:

```
length ["a","b","c"]
=     1 + length ["b","c"]
=     1 + 1 + length ["c"]
=     1 + 1 + 1 + length []
=     1 + 1 + 1 + 0
=     2 + 1 + 0
=     3 + 0
=     3
```

In each case the number of steps is approximately twice the number of items in the list, and the maximum space usage is approximately two symbols for each item of the list (to see this, rewrite the list as "*a*"::"*b*"::"*c*"::[]). So there is little to choose between the two versions in terms of efficiency.

In terms of simplicity, there is a big difference, though. The explicit version requires an inductive proof (or design using the method of differences), whereas the version using *foldl* requires only the invention of the $\oplus$ operator and the *i* element. So, for us

```
val length = foldl (fn (n,a) => n+1) 0;

val length = fn : 'a list -> int
```

An Example

Suppose we are given a list of student marks and asked to find the average. The marks are of type *int list*, the result will be of type *real*, so we are invited to design a function of type *int list* → *real*. The informal definition of the function is easy; we just add up the marks and divide by the number of students (which will be the length of the list). We already have a function to give us the length of a list; and the adding-up part looks like a repetitive operation, so we can use folding. There is a small problem of type conversion, as both the sum and the length will be integers; but we will separate that problem out (divide and conquer) and worry about it later.

Let's concentrate on adding up the marks. What will the $\oplus$ operator and the *i* element be? The *i* element is pretty obviously zero (no students, no marks); and the $\oplus$ operator looks like straight addition, which is fully associative, so we can use *foldl*, and obtain the following function for summing the list:

```
val sum_list = foldl op+ 0;

val sum_list = fn : int list -> int
```

(ML has been rather clever here. + is an overloaded operator, and therefore so is *op+*. But ML knows that the type of *foldl* is $('a * 'b \to 'a) \to 'a \to 'b\ list \to 'a$, where *'a* and *'b* are type variables. 0 is an integer, so *'a* must have value *int*. *op+* is of type $num * num \to num$, so *'a* and *'b* are the same type. It follows that both *'a* and *'b* have value *int* and the function *sum_list* is of type $int\ list \to int$. From the absolute minimum of clues, ML can successfully infer the type of the summing function.)

Now given a list *N* the sum will be *sum_list N* : *int* and the length will be *length N* : *int* as defined above. We need to convert these values to type real, and fortunately ML provides a function *real* to do this for us. Now we have the sum as *real*(*sum_list N*) : *real* and the length as *real*(*length N*) : *real*. All we need to do to obtain the average is to divide one by the other:

```
fun average N =
 real(sum_list N) / real(length N)
;

val average = fn : int list -> real
```

Now this definition, given *any* list of *any* length, will return the average mark. In designing it we didn't appeal to any principle of inductive proof (all that has been done earlier, and does not need to be done again), nor did we need to set up any complex structure to represent the repetition; everything was achieved by purely functional means.

Perhaps we had better run a few test cases, just to make sure:

```
average [1,2,3];

val it = 2.0 : real

average [40,50,60,70];

val it = 55.0 : real

average [];

ML EXCEPTION : Quot
```

Oh dear! We forgot about the null case. The *sum_list* function and the *length* function operate correctly for the null case, but when we combine them into the *average* function, we introduce the possibility of division by zero. We make a mental note to treat / and *div* with the respect they deserve.

Meanwhile, what should be the result of *average* []? The average does not exist, and so an exception is the appropriate response. The problem is that the name of the exception is not appropriate. Our rule for exceptions, argued for in Chapter 5, is that they should have the same name (with an initial capital) as the function in which they occur. So ideally we would like *average* [] to raise the exception *Average*.

One way to achieve this would be to add a special case to the definition of *average*:

```
exception Average
fun average [] = raise Average
|   average N  =
  real(sum_list N) / real(length N)
;

exception Average
val average = fn : int list -> real
```

This would ensure that the exception *Quot* can never be raised, as *N* would always be a non-null list. But this ploy involves an extra check on the argument of *average*, which in most cases is unnecessary. ML has a neater way of dealing with the situation, which is to **trap** the exception raised by division. The expression used to trap exceptions, called a **handle clause**, bears a strong similarity to the case-expression we met in Chapter 4:

```
exception Average
fun average N =
  real(sum_list N) / real(length N)
  handle Quot => raise Average
;

exception Average
val average = fn : int list -> real
```

The effect of the handle clause is to trap the *Quot* exception emanating from the evaluation of the expression *real*(*sum_list N*)/*real*(*length N*), and to raise instead the exception *Average*, which becomes the result of evaluating *average* []. The augmented expression *real*(*sum_list N*) / *real*(*length N*) handle *Quot* => raise *Average*

is a handle expression, which (like any other expression) may evaluate to give a result or an exception. As the handling mechanism is only brought into play when *Quot* is raised, this strategy is more efficient than the explicit checking of the null case we used earlier.

```
average [];

ML EXCEPTION: Average
```

The handle expression is quite general, and it would be perfectly in order syntactically to write

```
fun average N =
  real(sum_list N) / real(length N)
  handle Quot => 0.0
;

val average = fn : int list -> real
```

(provided, of course, that, as in this case, the type of the expression returned from the handle clause was compatible with that of the original expression). But semantically, this definition is faulty, as the average of a null list is not zero, it is undefined.

8.3 OTHER REPETITIVE OPERATIONS ON A LIST

We can use *foldl* and *foldr* to implement other repetitive operations on lists. For example the identity function takes a list

$$[a_1,a_2, \ldots a_{n-1},a_n]$$

and converts it into

$$[a_1,a_2, \ldots a_{n-1},a_n]$$

Not a very useful function, perhaps, but a good starting point. Can we rewrite the result in a form suitable for folding? Well, it could be expressed as

$$a_1 :: a_2 :: \ldots :: a_{n-1} :: a_n :: []$$

which is a right folding with [] as the identity element and :: as the ⊕ operator. So we can say

```
val I_list = foldr (op ::) [];

val I_list = fn : 'a list -> 'a list

I_list [1,2,3];

val it = [1, 2, 3] : int list
```

Mapping

The function *map* we met early on in this chapter is another useful repetitive function on lists. Given a function $g : 'a \to 'b$ and a list $A : 'a\ list$ it produces a $B : 'b\ list$ by applying g to each item of A. We can consider this to be a folding operation in which we convert

$$[a_1,a_2, \ldots a_{n-1},a_n]$$

to

$$g\ a_1 :: g\ a_2 :: \ldots :: g\ a_{n-1} :: g\ a_n :: []$$

In other words

```
fun map g = foldr (fn (a,B) => g a :: B) [];

val map = fn : ('a -> 'b) -> 'a list -> 'b list
```

A quick check:

```
map (fn x => 2*x) [1,2,3];

val it = [2,4,6] : int list
```

Can we make this function more efficient by using *foldl* instead of *foldr*? There is a **duality theorem** that can help us here. We have already seen that:

$$foldl\ \oplus\ i\ A\ =\ foldr\ \oplus\ i\ A$$

where ⊕ is a fully associative operator. But even when we don't have a fully associative operator, we can still have a choice of folding function if there are two operators ⊕ and ⊗ such that

$$a \oplus b \;=\; b \otimes a$$

because then

$$foldr \oplus i\, A \;=\; foldl \otimes i\, (rev\ A)$$

where *rev* is the function that reverses a list.

In our particular case the $\oplus$ operator is

$$(\text{fn } (a,A) \Rightarrow g\ a :: A)$$

so the associated $\otimes$ operator is the same function with the arguments reversed:

$$(\text{fn } (A,a) \Rightarrow g\ a :: A)$$

and we have the more efficient implementation

```
fun map g A =
  foldl (fn (B,a) => g a :: B) [] (rev A);

val map = fn : ('a -> 'b) -> 'a list -> 'b list
```

```
map (fn x => 2*x) [1,2,3];

val it = [2, 4, 6] : int list
```

(In fact, *map* is one of the standard functions provided by the ML environment, for this very reason of efficiency, but it is interesting to see that a reasonably efficient version can be defined by folding.)

Equivalence of *map* Implementations

Are the above definitions of map truly equivalent? The sceptical reader may not be satisfied by a simple test. It is, however, quite easy to prove that the folding definition of *map* is equivalent to the original one. The proof, as for all proofs of equivalence involving lists, is by induction. (I choose the inefficient right-folding version of *map* for simplicity, but the proof would be no more complicated in principle for the left-folding version.)

Proof We have to show that the functions defined by

```
fun map1 f (a::A)  = f a :: map1 f A
|   map1 f []      = []
;
```

and

```
fun map2 g = foldr (fn (a,A) => g a :: A) [];
```

are equivalent.

Base case

map1 ff []

= [] — definition of *map1*

= *foldr* (*fn* (*a*,*A*) => *ff a*::*A*) [] [] — definition of *foldr*

= *map2 ff* [] — definition of *map2*

Inductive case

Assume *map1 ff A* = *map2 ff A* for all *A* for which $0 \le length\ A \le j$

Now *map2 ff* (*a*::*A*)

= *foldr* (*fn* (*a*, *A*) => *ff a* :: *A*) [] (*a*::*A*) — definition of *map2*

= (*fn* (*a*, *A*) => *ff a* :: *A*) (*a*, *foldr* (*fn* (*a*, *A*) => *ff a* :: *A*) [] *A*) — definition of *foldr*

= (*fn* (*a*, *A*) => *ff a* :: *A*) (*a*, *map2 ff A*) — definition of *map2*

= *ff a* :: *map2 ff A* — function application

= *ff a* :: *map1 ff A* — inductive assumption

= *map1 ff* (*a*::*A*) — definition of *map1*

We have now shown that

map1 ff A = *map2 ff A* for all *A* for which $0 \le length\ A \le j+1$

QED

An Example

Suppose we are running a company and we identify our customers by a unique customer number. Our computer system has produced a list of invoices, each consisting of customer number and amount owing, and we want to convert these to a user-friendlier form by adding the customer name. So an invoice such as

(0300567, 268.75)

becomes

(0300567, 'Arkansas Aircraft Company', 268.75)

We could model the relation between customer numbers and customer names (which is many-to-one) as a function:

```
val customer_name =
  fn 0300567 => "Arkansas Aircraft Company"
  |  0679555 => "Bolton Brewing Corporation"
  |  _       => "UNKNOWN CUSTOMER"
;

val customer_name = fn : int -> string
```

Now the action which has to be performed on each invoice can be expressed as

```
val g = fn (a,b) => (a,customer_name a,b);

val g = fn : int * 'a -> int * string * 'a
```

and the overall operation on the invoices is

```
val friendly = map g;

val friendly = fn : (int * 'a) list ->
(int * string * 'a) list

friendly [(0300567, 268.75),(0679555,345.50)];

val it =
 [(300567, "Arkansas Aircraft Company", 268.75),
  (679555, "Bolton Brewing Corporation", 345.5)]
  : (int * string * real) list
```

The reader familiar with imperative programming may dismiss this example as totally unrealistic. Where are the layouts for the invoices? Where is the calculation of disk volume for the invoice file? How is the file to be read from disk? How are the invoices to be printed? Well, these are important aspects of the problem which we have chosen to ignore for the moment. We have defined the essence of the process, and everything else can be added in later using functional composition. We shall see how to deal with these aspects of the problem in Chapter 11.

Another objection to the above design is that it uses a function to hold data (the customer name database). In imperative programming this would be considered eccentric. In functional programing it raises no problems because functions can be used as arguments and results. It is perfectly possible, for instance, to write an updating function for the customer database:

```
fun update_database (n,s) f =
  fn m => if m = n then s else f m
;

val update_database = fn : ''a * 'b ->
(''a -> 'b) -> ''a -> 'b
```

This function takes a new entry (*n,s*) in the customer database, consisting of a customer number and a customer name, and the old customer database *f*, and returns the new customer database.

```
val new_customer_name =
  update_database
  (0575687,"Crawley Cornflakes Inc")
  customer_name;

val new_customer_name = fn : int -> string

new_customer_name 0575687;

val it = "Crawley Cornflakes Inc" : string

new_customer_name 0679555;

val it = "Bolton Brewing Corporation" : string

new_customer_name 0679556;

val it = "UNKNOWN CUSTOMER" : string
```

Filtering

Folding seems to be a very general form of operation on lists. Let's see if we can design a folded version of the function *filter* which we saw defined explicitly at the beginning of the chapter:

```
fun filter f (a::A) = if f a then a :: filter f A
                                  else filter f A
|   filter f []     = []
;

val filter = fn : ('a -> bool) -> 'a list -> 'a list
```

This is a function which converts

$$[a_1, a_2, a_3, a_4, \ldots a_{n-2}, a_{n-1}, a_n]$$

to some reduced list such as

$$[a_1, a_2, a_4, \ldots a_{n-2}, a_n]$$

dependent on the result of applying some Boolean function f to the items of the list. The i element is clearly [], but what is the $\oplus$ operator in this case? Rewriting the reduced list as

$$a_1 :: a_2 :: a_4 :: \ldots :: a_{n-2} :: a_n :: []$$

we see that it is an operator that sometimes prefixes an item to a list, and sometimes just returns the list, depending on the result of applying f to the item. This gives us an ML function g, corresponding to $\oplus$:

```
fun g f (a,A) = if f a then a::A else A;

val g = fn : ('a -> bool) -> 'a * 'a list -> 'a list
```

and a definition of *filter*:

```
local
 fun g f (a,A) = if f a then a::A else A
in
 fun filter f = foldr (g f) []
end;
```

or, more compactly:

```
fun filter f =
  foldr (fn (a,A) => if f a then a::A else A) []
;

val filter = fn : ('a -> bool) -> 'a list -> 'a list
```

Note that in this last example, by the rules of ML, the pattern f on the left-hand side of this definition binds the occurrence of f in the anonymous function definition on the right-hand side, in the absence of any intervening pattern f. This 'contextual' definition of an anonymous function can often be used to give compact definitions.

We could make *filter* more efficient, taking advantage of duality to design a left-folding version:

```
fun filter f A =
  foldl (fn (B,b) => if f b then b::B else B)
        []
        (rev A)
;

val filter = fn : ('a -> bool) -> 'a list -> 'a list
```

Here I have systematically renamed the variables bound by the fn expression. Although this is not strictly necessary (the scope rules of ML preclude any ambiguity) it avoids confusion in the mind of the reader.

Another Example

We want to buy an American motorhome, but have only £35 000 to spend. Given a list of vehicles and prices, we want to select only those in our price range. The list of vehicles and (sterling) prices will look something like:

```
[
("Amazon 20 RK",              33771),
("Eldorado L210/T255",        33850),
("Four winds 21-25 ft",       33852),
("Ultra Mini 1100",           35076),
("Sun Sport 22-32 ft",        42792),
("Meadowbrook 25-30ft",       43000),
("Vision IB24",               29995),
("Pinnacle 285DB",            54121),
("Allegro 23-28ft",           35244),
("Minnie Winnie 22-28ft",     34950)
] : (string * int) list
```

and the result will be a list of the same type, so we are looking for a function *select_motorhome* of type (*string* * *int*) *list* → (*string* * *int*) *list*. We could generalize the function to select for different price maxima (useful for our *next* motorhome); then the type would be

$$select\_motorhome\_up\_to : int \to (string * int)\ list \to (string * int)\ list$$

This function is clearly a filtering operation representable by *filter f*, in which the function *f* is one that returns true if the price of the motorhome is less than or equal to the amount we have to spend. That is:

```
fun f (n:int) (s,m) = m <= n;

val f = fn : int -> 'a * int -> bool
```

where *n* is the amount we have to spend and *m* is the price of the motorhome. (The type constraint is required because <= is overloaded.) So we have:

```
local
  fun f (n:int) (s,m) = m <= n
in
  fun select_motorhome_up_to n = filter (f n)
end;
```

or, more compactly:

```
fun select_motorhome_up_to (n:int) =
  filter (fn (s,m) => m <= n)
;

val select_motorhome_up_to = fn : int ->
('a * int) list -> ('a * int) list
```

and our required result is given by

```
select_motorhome_up_to 35000
[
 ("Amazon 20 RK",             33771),
 ("Eldorado L210/T255",       33850),
 ("Four winds 21-25 ft",      33852),
 ("Ultra Mini 1100",          35076),
 ("Sun Sport 22-32 ft",       42792),
 ("Meadowbrook 25-30ft",      43000),
 ("Vision IB24",              29995),
 ("Pinnacle 285DB",           54121),
 ("Allegro 23-28ft",          35244),
 ("Minnie Winnie 22-28ft", 34950)
];
```

```
val it =
 [("Amazon 20 RK", 33771),
  ("Eldorado L210/T255", 33850),
  ("Four winds 21-25 ft", 33852),
  ("Vision IB24", 29995),
  ("Minnie Winnie 22-28ft",34950)]
  : (string * int) list
```

from which we can select the motorhome of our dreams.

8.4 FINDING

It may appear that folding is the answer to all our list-processing problems, but there are certain kinds of function on lists that do not fit into the folding mould. Various kinds of searching function, for instance, do not need to scan the whole of the list.

Suppose we wish to find the first capital letter in a piece of text. We would first have to explode the text string to a list. We then have the more general problem of finding the first item in a list which satisfies a certain condition. This process will be modelled by a function taking a *'a list* and returning an item of *'a*. What if there is no item in the list which satisfies the condition? In this case we can raise an exception; we define 'finding the first item in a list' as a partial function.

We can specify the condition in the same way as we did for the *filter* function; as an argument which is itself a function of type $'a \rightarrow bool$. This gives us a function of the following type:

$$find\_first : ('a \rightarrow bool) \rightarrow 'a\ list \rightarrow 'a$$

We could certainly implement this as a species of filter, where we return only the first item of the filtered list:

```
exception Find_first
fun find_first f A =
  case filter f A
    of (b::B) => b
    |  []     => raise Find_first
;

val find_first = fn : ('a -> bool) -> 'a list -> 'a
```

but this would be rather inefficient if, for example, we were searching for the first capital letter in the *Complete Works of Shakespeare*. As many searching problems are of this type, with an average search length much less than the length of the list being searched, it will be advantageous to write an explicit search function which stops the scan after it finds the required item.

We can build such a function by the Method of Differences. The null case is straightforward:

$$\mathit{find\_first}\ f\ [\,] = \mathit{raise\ Find\_first}$$

and the inductive case is not hard either. As we increase the size of the list by prefixing an item, then *either* this item satisfies f and we have our answer *or* we have to search the rest of the list:

$$\mathit{find\_first}\ f\ (a::A) = \text{if } f\ a \text{ then } a \text{ else } \mathit{find\_first}\ f\ A$$

giving the ML code:

```
fun find_first f (a::A) =
  if f a then a else find_first f A
|   find_first f []      = raise Find_first
;

val find_first = fn : ('a -> bool) -> 'a list -> 'a
```

This function looks encouragingly efficient, as it is tail recursive. Now we can specialize *find_first* to search for the first capital letter in a string:

```
fun first_capital s =
  find_first is_uppercase (explode s)
;

val first_capital = fn : string -> string

first_capital "hello";

ML EXCEPTION: Find_first

first_capital "the lake of St. Wolfgang is one\
\ of the most beautiful in the Salzburg region";

val it = "S" : string
```

Yes, I agree that using the *explode* function on the *Complete Works of Shakespeare* is also inefficient, but there's not much we can do about that with our current knowledge of ML. Later on, in Chapter 10, we will see how to analyse long strings of characters efficiently.

A General Search Function

Our *find_first* function is still not as general as it could be. It will not give the answer to a question such as 'What is the position of the first capital letter in the *Complete Works of Shakespeare*?' Never mind for the moment whether this is a sensible question, let's try to generalize the *find_first* function to give us this information. We will require a function that returns a pair of values: the position in the list (of type *int*) and the value of the item (of type *'a*). The function will be partial as before, and will have the type

$$('a \to bool) \to 'a\ list \to int * 'a$$

Let's call the function *search* to emphasize its generality. The base case will be as before:

$$search\ f\ [] = raise\ Search$$

but the inductive case has a new wrinkle:

$$search\ f\ (a::A) = \text{if } f\ a \text{ then } (1,a) \text{ else } (n+1, a') \\ \text{where } (n,a') = search\ f\ A$$

We know from experience that such a specification will lead to an ML function definition with a *let* clause — a rather messy prospect. Under our motto *review and improve*, let's see if we can't turn this into a nice tail-recursive definition. After all, as we go through the list, all we're doing is incrementing the position by one for each item, and either rejecting or (finally) accepting the item. A standard strategy for making a function tail recursive is to add a new argument whose sole function is to carry the latest version of one of the result values. In our case, the value we wish to carry is the current position in the list, so we add this to our function specification:

$$search\ n\ f\ [] = raise\ Search \\ search\ n\ f\ (a::A) = \text{if } f\ a \text{ then } (n,a) \text{ else } search\ (n+1)\ f\ A$$

The inductive case is stating the indubitable fact that searching a list $a::A$, where the items are numbered from n, is going to result in *either* item n being returned *or* a result equivalent to that obtained by searching the list A where the items are number from n+1. This gives us the ML formulation:

```
exception Search
fun search n f (a::A) =
  if f a then (n,a) else search (n+1) f A
|   search n f []      = raise Search
;

exception Search
val search = fn : int -> ('a -> bool) -> 'a list
-> int * 'a
```

This new version of *search* is more general than the original one; we could number the items in the list from 1 or 0 or 7456 if we wanted to. It is also more efficient, as it is tail recursive. Normally in computing generality and efficiency are opposites which we have to trade off against one another. How have we achieved the miracle of improving both simultaneously?

The paradox disappears when we realize that generality is a property of a definition, while efficiency is a property of an implementation or evaluation. In conventional imperative computer languages, evaluation closely follows definition, so generality and efficiency are traded off similarly in both. But in functional computing languages, evaluation is quite different from definition, and in fact in recursive definitions, the evaluation process is the reverse of the definition process. So by not specifying the starting value of the position in the list when we define the function (which makes the definition more general), we force the starting value to be defined right at the beginning of the evaluation (which makes the evaluation more efficient).

We can now redefine *find_first* in terms of *search*, but before doing that we define two useful functions:

```
fun fst (a,b) = a;

fun snd (a,b) = b;

val fst = fn : 'a * 'b -> 'a
val snd = fn : 'a * 'b -> 'b
```

These functions return, respectively, the first and second element of a pair. They are called **selection functions**, or sometimes **projection functions** by analogy with projecting a figure onto an axis in analytical geometry, as they pull out one ‘coordinate’ of the pair.

Now function *find_first* is easy:

```
exception Find_first
fun find_first f A =
  snd (search 1 f A)
  handle Search => raise Find_first
;

val find_first = fn : ('a -> bool) -> 'a list -> 'a
```

You can see that in this case, the definition without the selection function would be much clumsier:

```
exception Find_first
fun find_first f A =
  let
    val (n,a) = (search 1 f A)
  in
    a
  end
  handle Search => raise Find_first
;

exception Find_first
val find_first = fn : ('a -> bool) -> 'a list -> 'a
```

We can also define a function, *position*, which will give the position in the list of the first item satisfying an arbitrary condition:

```
exception Position
fun position f A =
  fst (search 1 f A)
  handle Search => raise Position
;

exception Position
val position = fn : ('a -> bool) -> 'a list -> int
```

and it would be equally easy to define *position_0* which gives the position counting from zero.

Other Search Functions

Before we leave search functions, there are two other very general functions which we can derive from our *find_first* function. These are functions which answer the questions:

- Are there any items with a certain property in the list?
- Do all the items in a list have a certain property?

These functions are called *any* and *all*. Here are some examples:

$$any\ (\text{fn}\ n \Rightarrow n\ div\ 2 = 0)\ \ [1,2,3,4] \ = \ true$$

$$all\ (\text{fn}\ n \Rightarrow n\ div\ 2 = 0)\ \ [1,2,3,4] \ = \ false$$

any is a pretty obvious specialization of *find_first* which returns *true* if *find_first* finds anything, and false if it doesn't (that is, if it raises an exception):

```
fun any f A =
  let
    val a = find_first f A
  in
    true
  end
  handle Find_first => false
;

val any = fn : ('a -> bool) -> 'a list -> bool
```

The exception handling makes this code look a bit tricky, so let's analyse it in detail. The value of the expression *any f A* is defined to be

$$\text{let val}\ a = find\_first\ f\ A\ \ \text{in}\ true\ \text{end}$$

This expression will evaluate to two possible results, the value *true* if *find_first* finds an item in *A* satisfying *f*, and the exception *Find_first* otherwise. (The identifier *a* is evaluated but discarded.) If the exception is raised, the handle clause will convert it to value *false*, thus giving the correct result in all cases.

The function *all* is less obviously a derivative of *find_first*, in fact it looks more like a derivative of *filter*. But wait! All the items in a list satisfy *f*, if and only if there are no items in the list which satisfy *not f*.

```
fun all f A = not (any (not o f) A);

val all = fn : ('a -> bool) -> 'a list -> bool
```

As this function gives the answer *false* as soon as it finds an item not satisfying *f*, it is an efficient implementation of *all*. It also has the consequence that

$$all\, f\, [] \;=\; true$$

for any *f* whatsoever. This rather surprising result can be reconciled with our intuition by noting that it is impossible to find an item in [] for which the function *f* will give the result *false*.

One more search function that is worth mentioning is called *is_item*. It returns *true* if and only if it can find a value *a* in a list *A*. It is a specialization of *any*:

```
fun is_item a = any (fn a' => a' = a);

val is_item = fn : ''a -> ''a list -> bool
```

(Note that the list items must be equality types.) So, for example, we have:

```
is_item 3 [1,2,3];

val it = true : bool

is_item 4 [1,2,3];

val it = false : bool
```

8.5 COMBINING LISTS

All the functions we have looked at so far operate on a single list; they are similar to unary operations like *sin* or *sqrt*. What about combining several lists? Are there functions on lists which correspond to binary operations like + or *mod*? The answer is: yes, there are, and just as in the case of simple variables, we can build up expressions of as much complexity as we wish by the sequential application of binary functions to lists.

Once again, as for folding and searching, we will derive the general function for combining lists, and then specialize it in various ways. We can use a simple binary function such as + as a guide — it takes a pair of numbers (of the same type) as argument and returns a number of the same type as result.

$36 + 62 = 98$

The equivalent function for lists will take a pair of lists containing numbers of the same type, and return a single list containing the sums of corresponding items as a result.

[36,	list_+	[62, =	[98,
86,		44,	130,
23,		12,	35
90]		~77]	13]

Mathematicians will recognize this operation as **vector addition**.

Of course, we could use any binary function instead of +; it could be a completely general function taking two different types and returning a third, say $\otimes : 'a * 'b \rightarrow 'c$. Then our general *combine* function will have the type

$$combine : ('a * 'b \rightarrow 'c)\ ('a\ list * 'b\ list) \rightarrow 'c\ list$$

(I have chosen to make the second argument a pair of lists rather than to curry the function completely; this will make it possible to define infix list functions later.)

When we think about defining our general *combine* function, the method of differences gives us the inductive case easily: suppose

$$combine \otimes (A, B) = C$$

then increasing the size of the lists by one produces

$$combine \otimes ((a{::}A),(b{::}B)) \ = \ (a \otimes b) :: C$$

giving the recurrence relation

$$combine \otimes ((a{::}A),(b{::}B)) \ = \ (a \otimes b) :: combine \otimes (A, B)$$

The null case is straightforward too:

$$combine \otimes [\,]\,[\,] \ = \ [\,]$$

Translating these into ML-ese, remembering that only prefix functions are allowed as arguments, we obtain:

```
fun combine f ((a::A),(b::B)) =
    (f(a,b))::combine f (A,B)
|   combine f ([]    ,[])     = []
;

ML WARNING — Clauses of function binding are
non-exhaustive
INVOLVING:  fun combine
val combine = fn : ('a * 'b -> 'c) -> 'a list *
'b list -> 'c list
```

ML warns us that we have not defined all the possible cases of this function. Which cases have we missed out? Just those where the two lists *A* and *B* have different lengths. But in these cases the result of the function is not well defined, so we must accept that our *combine* function is partial, and raise an exception whenever *A* and *B* are not the same length:

```
exception Combine
fun combine f ((a::A),(b::B)) =
    (f(a,b))::combine f (A,B)
|   combine f ([]    ,[])     = []
|   combine f (_)             = raise Combine
;

exception Combine
val combine = fn : ('a * 'b -> 'c) -> 'a list *
'b list -> 'c list
```

Now we have our combine function, we can easily define vector addition on integer lists:

```
infix 6 v_add
val (v_add: int list * int list -> int list) =
    combine op+
;

infix 6 v_add
val op v_add = fn : int list * int list -> int list
```

I have given *v_add* the same precedence as normal addition, and this is the general rule for these list versions of binary operations. In theory we could produce versions of all the standard ML binary operators, and duplicate the standard precedences exactly.

```
[36,86,23,90]  v_add [62,44,12,~77] ;
val it = [98, 130, 35, 13] : int list

[36,86,23,90]  v_add [62,44,12] ;
ML EXCEPTION: Combine
```

8.6 RELATIONS BETWEEN FUNCTIONS

We have now defined some eighteen functions in this chapter. Four of the functions, *foldl*, *foldr*, *search* and *combine*, have been defined explicitly using the Method of Differences. All the other functions have been derived from these four. The proofs for these derived functions do not involve the use of induction at all. In this way a tremendous amount of sweat and tears has been avoided, and the efficiency which we have built into the functions *foldl* and *search* has been bequeathed to their derivatives.

Figure 8.1 shows the relationships between the functions we have defined so far. It is fascinating to see the ease with which extremely general functions like *foldr* can be tailored to give useful specializations such as *select_motorhome_up_to*.

8.7 ODDS AND ENDS

Despite our success in specializing functions, there are a few functions which cannot be fitted into any general pattern. For example, suppose we want to check whether precisely *one* item in a list has a certain property,

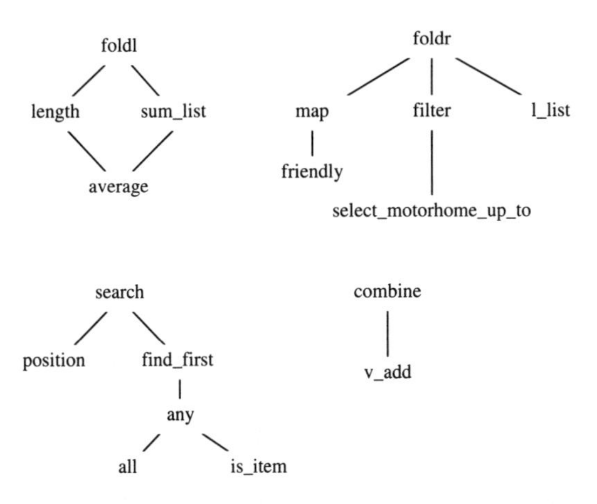

Fig. 8.1 Relations between functions

rather than *any* number of items. At first blush, this function looks like a derivative of *find_first*. But it is not, and the reason is as follows. To verify that one and only one item has the required property, we cannot stop when we find that item, we have to continue until another such item is found or we reach the end of the list. We could, of course, always scan the entire list using *filter*:

```
fun one f A = length (filter f A) = 1;

val one = fn : ('a -> bool) -> 'a list -> bool
```

but the Shakespeare effect hits us again, to show that

one (fn *s* => *s* = "a") *complete_works_of_shakespeare* = *false*

would require a complete scan. A better option is to design function *one* from scratch using the Method of Differences. We have

```
one f [] = false
```

for the base case, and for the inductive case *one f* (*a*::*A*) = *true* if either *f a* is true and *any f A* is false, or *f a* is false and *one f A* is true. Formulating this in ML:

```
one f (a::A) =
  if f a then not (any f A) else one f A
```

giving the complete function

```
fun one f (a::A) =
  if f a then not (any f A) else one f A
|   one f []       = false
;

val one = fn : ('a -> bool) -> 'a list -> bool
```

Because the ML if-expression is implemented in terms of pattern-matching, this formulation executes quite sweetly. The evaluation of *one f* (*a*::*A*) causes *f a* to be evaluated; if this proves false then *one f A* is evaluated, and so on until the null list is reached. If, however, *f a* is true, then *not* (*any f A*) is evaluated, a process that in general will not require a complete scan of *A*. In this way the Shakespeare effect is avoided.

```
one (fn n => n = 1) [1,2,3,4];
```

```
val it = true : bool
```

```
one (fn n => n = 1) [1,2,1,3,4];
```

```
val it = false : bool
```

```
one (fn n => n = 1) [2,3,4];
```

```
val it = false : bool
```

The nth Item of a List

Finding the nth item of a list is another operation that doesn't fit into the folding or searching mould. We can imagine a function *nth* such that *nth k A* gives the kth item in list *A*. It might be used, for example, in a text-encrypting program. One way of encrypting text is to use a substitution cypher in which each character of the text is is substituted by another (fixed) character. Although such a cypher is fairly easy to crack (by considering letter frequencies), it looks impressive. A functional programmer would tend to define a function to do the substitution:

```
val encrypt = fn "a" => "g"
               | "b" => "r"
               | "c" => "d"
               ...
```

but as we might want to generate the encryption table automatically in some way, there is a case for using a list. As the ASCII table defines a numerical code for each letter of the alphabet, it is easy to convert letters 'a' ... 'z' to numbers 1 ... 26. If we have at hand a list of the 26 target letters:

```
val target = ["g","r","d",...];
```

it is possible to implement *encrypt* using *nth*:

```
fun encrypt s = nth (ord s – ord "a" + 1) target;
```

The question is: how do we implement *nth*? It turns out that the best way is the most straightforward — use the Method of Differences. We can obtain an expression for *nth k* (*a*::*A*) in terms of *A* very easily. If *k* is 1, the answer is *a*. Otherwise, the kth entry in *a*::*A* is the *k*-1th entry in *A*:

```
nth 1 (a::A) = a
nth k (a::A) = nth (k-1) A
```

The only remaining problem is the null case. This is undefined for all k, so we have

```
nth k [] = raise Nth
```

Putting it all together, we obtain:

```
exception Nth
fun nth 1 (a::A) = a
|   nth k (a::A) = nth (k-1) A
|   nth k []     = raise Nth
;

exception Nth
val nth = fn : int -> 'a list -> 'a
```

There is an interesting point about pattern-matching and efficiency here. The null case is very unlikely to arise in practice, and so is put last. The $k > 1$ case is more likely to occur than the $k = 1$ case, but because it includes the $k = 1$ case, it has to be placed after it to avoid a mismatch. This is a general rule about pattern-matching. A pattern which includes another pattern must be placed after it in the list of match rules. Patterns which are non-overlapping should be listed in order of decreasing likelihood.

We have omitted considering the case where k is negative, and at first glance it seems that our function is only partial. What happens when k is negative?

The answer is: exactly the same as when k is greater than the length of the list — an exception is raised, because the null list is bound to be reached eventually in the process of evaluation.

8.8 SYNTAX INTRODUCED IN THIS CHAPTER

exp	::=	*exp* handle *match*	handle expression
match	::=	*mrule* ⟨ | *match*⟩	
mrule	::=	*pat* => *exp*	
atpat	::=	*excon*	exception constant

8.9 CHAPTER SUMMARY

List processing constitutes a large part of functional programming. A vast number of operations on lists can be expressed in terms of the higher-order concepts of folding, searching and combining.

Folding is the process of combining all the items in the list using some binary function, conventionally thought of as an infix operator $\oplus$. If $\oplus$ is right associative, folding must be done from the right, if $\oplus$ is left associative, folding must be done from the left. Fully associative operators such as + can be folded either way, but left folding is preferred on grounds of efficiency, as it is tail recursive.

Useful specializations of folding include **mapping** one list to another, and **filtering** a list for all items satisfying a certain condition.

Searching is the process of finding the first item in a list which satisfies certain criteria. Although, on average, this takes time proportional to the length of the list, searching can be quite efficient, as the search stops once an item is found.

Combining is the process of taking corresponding items of two lists, and applying a binary operator (say $\otimes$) to each pair of items to produce a new list.

Some operations on lists do not fit the folding, searching or combining mould, and have to be written explicitly using the Method of Differences.

EXERCISES

Design and test functions to perform the following tasks. Try to re-use functions which have been defined already.

1. A function *discount* : *real list* → *real list* which takes a list of prices and produces a list with 10% discounted from the prices. What happens if the original list contains negative numbers?
2. A function *passes* : (*string* * *int*) *list* → (*string* * *int*) *list* which takes a list of student names and marks (as integer percentages) and returns a list of all those who have achieved 40% or more.

3. A function *flat* : *'a list list* → *'a list* which takes a list of lists and 'flattens' it into a list. So

   flat [[1,2,3], [4,5], [6], []] = [1,2,3,4,5,6]

4. In Synchronous Drive, the houses are numbered consecutively, starting from one. Each house has a name, the list of names (in order) being as follows:

```
val synchronous_drive =
[
  "Infinite Stream",
  "Graceful Degradation",
  "Stable Platform",
  "Unbundled",
  "Dynamic Ram",
  "Windows Three",
  "Odd Parity"
];
```

   Write a function *house_name* : string list → *int* → *string* that takes the above list and the number of a house on synchronous drive as arguments and returns the name of the house, also a function *house_number* : string list → *string* → *int* that takes the above list and a name as argument and returns the number of the house.
5. A function *mark_of* : (*string* * *int*) *list* → *string* → *int* which takes the list of students and marks of Exercise 2, and a student name, and returns that student's mark.
6. An infix function *zip* : *'a list* * *'b list* → (*'a* * *'b*) *list* which 'zips' a pair of lists together into a list of pairs. Give it precedence 5, to allow equality (precedence 4) to be applied to expressions involving zipped lists.
7. And its inverse, *unzip* : (*'a* * *'b*) *list* → *'a list* * *'b list*. Check they are true inverses by running a test.
8. Design and test functions *max* : *int list* → *int* and *min* : *int list* → *int* which return the highest and lowest values, respectively, in a non-null list.
9. Design and test a function

   new_nth : *int* → *'a* → *'a list* → *'a list*

   such that *new_nth n a A* gives a new list A', which is A with its nth item replaced by *a*. Use *nth* as a guide.

10. Design and test the function *take* which takes an integer n and a list A, and returns the first n items of the list. Do sensible things if n is too large or too small.
11. Design and test the complementary function *drop* which returns all but the first n items of a list. Again, do sensible things on n out of range.
12. Prove by induction that

    $$\textit{take}\ n\ A\ @\ \textit{drop}\ n\ A\ =\ A$$

    for all n and A. If it doesn't, change your definition of *take* and *drop*!

CHAPTER

NINE

CHARACTER LISTS

9.1 INTRODUCTION

In Chapter 8 we studied all kinds of operations on lists. This study is about to be brought to good use as we look at character lists. We make a particular point of examining character lists in some detail, as sequences of characters represent the only way (in ML anyway) in which the user can communicate with the computer. This may seem a strong restriction on human–computer interaction, but it is not. All kinds of interactions using a keyboard and mouse can be modelled as a sequence of characters; likewise most screen painting can be represented in terms of a sequence of characters.

We shall look at character lists in some detail, and concentrate on some of the type conversion problems which occur in human–computer interaction.

9.2 BOOLEAN FUNCTIONS ON CHARACTERS

To begin with, we consider some of the simplest problems we can encounter when dealing with character input by a user; deciding whether we have a digit, a letter or some other character. All these problems involve determining whether a character is within a certain range of values in the ASCII table. We have already designed a function to tell us whether a number is in the range represented by two other numbers (Chapter 6, Exercise 7):

```
fun is_in_range i j (n:int) =
  i <= n andalso n <= j;

val is_in_range = fn : int -> int -> int -> bool
```

(Note that the type-cast is necessary because <= is overloaded on integers and reals.)

We can now specialize this function to test whether a character lies between two other characters:

```
fun char_is_in_range s t u =
  is_in_range (ord s) (ord t) (ord u)
;

val char_is_in_range = fn : string -> string ->
string -> bool
```

As usual, we must reckon with ML's habit of treating characters as strings, which means that it is perfectly good ML to write

```
char_is_in_range "aardvark" "zebra" "lion";

val it = true : bool
```

but (luckily) not good form to write

```
char_is_in_range "" "z" "l";

ML EXCEPTION: Ord
```

Having found the general form, the individual functions are easy to write:

```
val is_digit = char_is_in_range "0" "9";

val is_digit = fn : string -> bool

val is_lowercase = char_is_in_range "a" "z";

val is_lowercase = fn : string -> bool

val is_uppercase = char_is_in_range "A" "Z";

val is_uppercase = fn : string -> bool
```

and the reader could easily dream up more examples, I'm sure.

Boolean Functions on Character Lists

Now we move on to the equivalent functions on character lists, and immediately find a pattern we have met before in Chapter 8; testing every item in a list for a certain condition. This pattern of processing we christened the *all* function; it had type *('a → bool) → 'a list → bool* if you remember.

Using the *all* function, our character list functions become trivial (they are all of type *string_list → bool*):

```
val is_digit_cl = all is_digit;

val is_lowercase_cl = all is_lowercase;

val is_uppercase_cl = all is_uppercase;
```

and so on. (Because of the way we have defined *all*, note that a null list will give a true result for all these functions.)

9.3 JUSTIFICATION

Now let's turn to the output of character lists. A challenging problem would be to

left - justify,

right justify,

or centre

characters within a line, or more generally, within a character string of fixed length. This will involve adding space characters to the original characters, but how many, and where?

For left and right justification of a character list *cl* in a character string of length *n*, the answer is clear; we must add *n - length cl* space characters. Let's assume we can generate a list of anything by using the function *list_of: 'a → int → 'a list*; which you invented as an exercise in Chapter 6. Then we have

```
fun left_justify n cl =
  cl @ list_of " " (n-length cl)
;

val left_justify = fn : int -> string list ->
string list
```

and

```
fun right_justify n cl =
  list_of " " (n - length cl) @ cl
;

val right_justify = fn : int -> string list ->
string list
```

Centring requires a little more thought. We still need to add *n - length cl* spaces, but this time divided on either side of the original character list. If *n - lengthcl* is not even, we end up with two unequal lists of spaces, and we have to decide which one to put where. I propose the convention that we put the shorter list first (it usually looks better) and say

```
fun centre n cl =
let
  val space_length = n-length cl
  val length_1 = space_length div 2
  val length_2 = space_length - length_1
  val spaces = list_of " "
in
  spaces length_1 @ cl @ spaces length_2
end
;

val centre = fn : int -> string list -> string
list
```

Now for a few tests:

```
left_justify 20 ["h","e","l","l","o"];

val it = ["h", "e", "l", "l", "o", " ", " ",
" ", " ", " ", " ", " ", " ", " ", " ", " ", " ",
" ", " ", " "] : string list
```

Well, it works, but the results are not very pleasing to the eye. Although character lists are convenient from the computer's point of view, they are the last thing the user wants to see. We shall have to ensure that character lists are converted to a more readable form when they are presented to the user. For the moment, we use the built-in functions *implode* and *explode* from Chapter 4 to make our tests slightly more readable:

```
implode(left_justify 20 (explode("hello")));

val it = "hello               " : string
```

Normally, of course, we convert input from the user into a character list just once, immediately we receive it, and work in terms of character lists throughout, converting back to a readable form at the last possible moment before output. It is only for the purposes of our tests that we have to put in the explicit conversions:

```
implode(right_justify 20 (explode("hello")));

val it = "               hello" : string

implode(centre 20 (explode("hello")));

val it = "        hello       " : string
```

9.4 CONVERSION BETWEEN NUMBERS AND CHARACTER LISTS

We have decided to use character lists as our internal data medium, because they are convenient to process, but the other basic ML data types will be useful too, and therefore we need conversion functions between character lists and the ML basic types: *int*, *real*, *string* and *bool*. Most languages perform these kinds of conversions automatically, when data is input and output, and this is true of ML in interactive mode, as we have seen. But for the more complex, large-scale systems we now want to build, it will be advantageous to separate the process of input and output (using character lists) from the process of validating data and converting between types.

In this way we obtain finer control over the values the user is allowed to input; for instance, if we are presenting the user with a menu of choices, we may require the selection to be an integer in the range 1 to 6 in order to be valid. The ML type system on its own cannot give us this degree of control, and certainly cannot deal in a user-friendly way with the user entering an invalid reply. We shall deal with the important issue of validation later in the chapter; for the moment we concentrate on type conversion.

As we have seen, the standard functions *implode* and *explode* allow us to convert between character lists and strings, and the conversion between character lists and Booleans is straightforward, so we shall concentrate on converting between numbers and character strings.

Character Lists to Numbers

As usual, we shall proceed from the simple to the more complex. The simplest conversion we can imagine is from a character list to a natural number or positive integer. Like all conversions from character lists, it will be a partial function; there will be some values of character list which do not correspond to any natural number, and these will raise an exception. We can use our *is_digit_cl* function to eliminate these values, but because of its habit of counting [] as a valid list of digits, we will have to eliminate the null list separately:

```
exception Cl_pos_int
fun cl_pos_int [] = raise Cl_pos_int
|   cl_pos_int S  =
   if not (is_digit_cl S) then raise Cl_pos_int
. . .
```

Now we come to the heart of the function, which takes a list of decimal digits and converts them to an integer. A few examples:

```
["0"]           ->  0
["0","1"]       ->  1
["1","1"]       ->  11
["0","9","1"]   ->  91
```

Is there any rhyme or reason to the conversion? Let's ignore leading zeros for the moment, as they seem to bring difficulties, and try gradually increasing the number of characters in the list:

```
["1"]           ->  1
["1","2"]       ->  12
["1","2","3"]   ->  123
```

The pattern now becomes clear: at each stage, adding another digit to the end of the list causes the result to be multiplied by ten and the converted value of the new digit to be added in. Furthermore, this process will deal with leading zeros too. We have a left-folding operation

fn (i,s) => 10 * i + ord s - ord "0"

with an identity element corresponding to a list of all zeros, that is, 0. We now write down the function:

```
exception Cl_pos_int
fun cl_pos_int [] = raise Cl_pos_int
|   cl_pos_int S  =
  if not (is_digit_cl S) then raise Cl_pos_int
  else
    foldl (fn (i,s) => 10 * i + ord s - ord "0")
          0
          S
;
```

```
exception Cl_pos_int
val cl_pos_int = fn : string list -> int
```

```
cl_pos_int ["0","1","0","9"];
```

```
val it = 109 : int
```

```
cl_pos_int ["0","1","0","a"];
```

```
ML EXCEPTION: Cl_pos_int
```

Having converted positive numbers, it is an easy matter to deal with negative ones; they are simply positive numbers with a minus sign in front. And I mean a minus sign here; we are dealing with the user, who knows nothing of ML's strange conventions for negative numbers. We can invent a function called *cl_int* to convert a character list to an integer. The function will be partial, of course; the list has to be non-null, and the corresponding positive number has to be convertible by *cl_pos_int*:

```
exception Cl_int
fun cl_int [] = raise Cl_int
|   cl_int ("-"::S) = (~(cl_pos_int S)
      handle Cl_pos_int => raise Cl_int)
|   cl_int S = cl_pos_int S
      handle Cl_pos_int => raise Cl_int
;
```

```
exception Cl_int
val cl_int = fn : string list -> int
```

(The brackets around the first handle expression are necessary to avoid ML misinterpreting the | as part of the handle clause. Be careful with handle expressions, as they can lead to this sort of ambiguity.)

Testing this:

```
cl_int [];

ML EXCEPTION: Cl_int


cl_int ["-","2","3"];

val it = ~23 : int


cl_int["3","4"];

val it = 34 : int
```

Character lists representing real numbers can be converted in similar but slightly more complex ways; for details the reader is referred to Appendix 3.

Numbers to Character Lists

Now for the conversion in the opposite direction, and here we have the following pattern for positive numbers:

```
1        ->        ["1"]
12       ->        ["1","2"]
123      ->        ["1","2","3"]
```

At each stage, the last decimal digit of the number *n* gives the character to be appended to the list. We can obtain this decimal digit by taking the number modulo 10

```
val last_digit = n mod 10
```

and then convert it to the appropriate ASCII character by using the *chr* function:

```
val last_char = chr(last_digit + ord "0")
```

The remaining digits can be obtained by dividing the number by 10 and ignoring the remainder (our *div* function).

```
val front_digits = n div 10
```

Now our recurrence relation is

pos_int_cl n = *pos_int_cl* (*front_digits*) @ [*last_char*]

The base case for the induction occurs when the number to be converted has only one digit, that is, when *front_digits* = 0. Putting all this together we obtain

```
fun pos_int_cl n =
  let
    val last_digit = n mod 10
    val last_char = chr(last_digit + ord "0")
    val front_digits = n div 10
  in
    if front_digits = 0 then [last_char]
    else pos_int_cl front_digits @ [last_char]
  end
;

val pos_int_cl = fn : int -> string list

pos_int_cl 0;

val it = ["0"] : string list

pos_int_cl 123;

val it = ["1", "2", "3"] : string list
```

The function appears to work nicely, but under our motto of *review and improve* we may spot a better way to do the conversion. The nasty part of this function is the list concatenation required to append one character to the list. A prefixing operation would be so much more elegant and efficient. One way to obtain this efficiency is to hold the list being built up as an extra argument of the function, and prefix characters to it as they are removed from the back end of the number. Calling our modified (tail recursive) function *picl*, we have :

	picl	123	[]
=	*picl*	12	["3"]
=	*picl*	1	["2","3"]
=	["1","2","3"]		

The base case, as before, occurs for a single-digit number. We can express this new recurrence relation in ML as:

```
fun picl n S =
  let
    val last_digit = n mod 10
    val last_char = chr(last_digit + ord "0")
    val front_digits = n div 10
  in
    if front_digits = 0 then last_char::S
    else picl front_digits (last_char::S)
  end
;

val picl = fn : int -> string list -> string list
```

giving the revised function:

```
fun pos_int_cl n = picl n [];

val pos_int_cl = fn : int -> string list


pos_int_cl 0;

val it = ["0"] : string list

pos_int_cl 123;

val it = ["1", "2", "3"] : string list
```

It works too.

The extension of this idea to negative integers and to real numbers is straightforward, and again the reader is referred to Appendix 3.

9.5 VALIDATION

Now, as promised, we turn to validation, and we find that we have done the groundwork for it already in our treatment of conversion. First, let's discuss what we mean by validation. It can be seen as a conversation between the user and the computer, in which input data values are agreed. The computer has certain fixed criteria for the values to be input. The user has some ideas too, and these two views of the data may not agree. When this situation occurs, it is essential that the mutual feedback process between the two sides be as efficient as possible. In other words, we are looking for an immediate response from the computer whenever the user enters incorrect data, and some kind of re-try facility to allow the user to correct the data.

In a true conversation, of course, the user could persuade the computer that the original value was correct, but this level of sophistication is beyond us at present.

In any case, as part of the validation process, it is necessary that the computer can unambiguously decide whether a particular piece of data is valid or not. In ML this decision can be made as a by-product of the conversion process: if we can't convert the user's data into the form we want, it is invalid. A natural way to model validation, then, is to apply a conversion function to the data the user has input, and see if the conversion is successful. If it isn't, the function will return an exception.

A little more than this is needed, though, to obtain the fast feedback we require. It would be very helpful to the user if the computer explained exactly what was wrong with the data, so that the next attempt stands a better chance of success. We can use the more elaborate form of exception, first seen in the previous chapter, to achieve this goal. We design a general form of exception for validation errors, which takes a string as argument. When the exception is raised by a validation function, the string argument contains a helpful hint as to what might be wrong with the data. This string can be output to the user as advice for the re-try process. The exception can be declared as follows:

```
exception Verr of string;
```

So, supposing we wish to validate that the user has entered an integer, the following function would suffice:

```
fun valid_integer S = cl_int S
  handle Cl_int =>
  raise Verr "not an integer number"
;

val valid_integer = fn : string list -> int
```

Example of a Validation Function

A full set of validation functions is given in Appendix 4, but as an exercise, let's invent a validation function for dates. Our dates will be six digits in the format DDMMYY, assuming the year to be in the range 1901 to 2099 (by this time, a better functional language than ML may have appeared). This convention means that two dates in two different centuries can have the same representation, but we don't worry about this — we just perform the validation and let the rest of the system work out which century the user is talking about. Because we're not trying to evaluate the date, our function, *valid_date*, will return its argument as result (if all goes well), so it will have the type *string list* → *string list*.

How shall we design this function? There are clearly complications with leap years, and the differing numbers of days in each month, and in any case, how do we analyse the date string to extract the days, months and years? The task seems overwhelming until we remember our golden rule, *divide and conquer*. If we design the function as a set of sub-functions, each dedicated to one part of the task, we may yet be successful.

First, then, let's look at splitting the date string into days, months and years. We have a six-character list, and we want to end up with three two-character lists:

```
fun split_date [c1,c2,c3,c4,c5,c6] =
 ([c1,c2],[c3,c4],[c5,c6])
;

ML WARNING - Clauses of function binding are
non-exhaustive
val split_date = fn : 'a list -> 'a list * 'a
list * 'a list
```

The warning message tells us that this is a partial function, so we add an exception:

```
exception Split_date
fun split_date [c1,c2,c3,c4,c5,c6] =
  ([c1,c2],[c3,c4],[c5,c6])
|   split_date _ = raise Split_date
;

exception Split_date
val split_date = fn : 'a list -> 'a list * 'a
list * 'a list

split_date ["1","2","3","4","5","6"];

val it = (["1", "2"], ["3", "4"], ["5", "6"]) :
string list * string list * string list
```

Now we need to check the ranges of the days, months and years. To do this, we need to convert them to integers. This is no problem as we have already invented *cl_int* to perform this task.

Now for the range checks themselves. As far as the years are concerned, there is no need for a check, assuming the conversion was successful. All values from 00 to 99 are valid. For the months we require a simple check:

```
fun is_month n = 1 <= n andalso n <= 12;

val is_month = fn : int -> bool
```

For days, the exact range depends on the month, and (for February) the year, so we need day, month and year as arguments:

```
fun is_day d m y =
  1 <= d andalso d <= num_of_days m y
;
```

This is a pleasant feature of design in a functional language; if the logic seems to become too complex, we can always invent a function to simplify things. Now, however, we need to write the function *num_of_days*. The rhyme we learnt as children will come in handy here:

```
fun num_of_days m y =
  case m of
    9 => 30
  | 4 => 30
  | 6 => 30
  |11 => 30
  | 2 => if is_leap y then 29 else 28
  | _ => 31
;
```

Once again we have invented a function, but this time it is easy to implement, as by great good fortune, the year 2000 is a leap year (the century number being divisible by four):

```
fun is_leap y = y mod 4 = 0;
```

(This is, of course, why we chose the range of years to be 1901 to 2099.) Now we have our *valid-date* function; it's just a question of putting the bits together:

```
fun valid_date S =
  let
    exception Split_date
    fun split_date [c1,c2,c3,c4,c5,c6] =
      ([c1,c2],[c3,c4],[c5,c6])
    |   split_date _ = raise Split_date
    ;
    fun is_month n = 1 <= n andalso n <= 12;
    fun is_leap y = y mod 4 = 0;
    fun num_of_days m y =
      case m of
        9 => 30
      | 4 => 30
      | 6 => 30
      |11 => 30
      | 2 => if is_leap y then 29 else 28
      | _ => 31
    ;
    fun is_day d m y =
      1 <= d andalso d <= num_of_days m y
    ;
    val (dd,mm,yy) = split_date S
      handle Split_date =>
      raise Verr "not of length 6"
    val d = cl_int dd
      handle Cl_int =>
      raise Verr "has invalid day"
    val m = cl_int mm
      handle Cl_int =>
      raise Verr "has invalid month"
    val y = cl_int yy
      handle Cl_int =>
      raise Verr "has invalid year"
in
  if not (is_month m)
  then raise Verr "has invalid month"
  else if not (is_day d m y)
  then raise Verr "has invalid day"
  else S
end;
```

```
val valid_date = fn : string list -> string list
```

Questions

1. How many tests would be necessary to check the logic of this function?
2. How would you modify the function to accept the American form of date (MMDDYY)?

Answers

1. I calculate 36, though some of them could be combined. The problem is testing the maximum days of the month, for all months and for normal and leap years. What a pity that the Emperor Augustus complicated matters so much.
2. Simple. Just modify *split_date*.

9.6 AN EXAMPLE OF CHARACTER LISTS IN ACTION: THE MATHEMATICAL GENIUS

To demonstrate some of the power of the operations on character lists, we will build a mathematical genius or *idiot savant*. This is a function of type *string* → *string* that will accept orders such as 'multiply 3268347 and 574839' and reply with 'Easy! The answer to your question is 1878773321133.' We'll allow our idiot savant to do addition and subtraction, multiplication and division.

It is clear that the main problem in building the idiot savant in ML is not performing arithmetic operations on large numbers, but recognizing a number, and recognizing words like 'multiply' and 'add'. Our genius will be completely baffled if we ask it 'How do you feel today?', and will reply with 'I don't understand your question'.

So our first task is to build a word recognizer that will split our input sentence into words. Later on we can write a function to recognize certain words as numbers or **keywords** like 'add'. Finally, we need to perform the arithmetic operations, and then convert the result back into a sentence. Each one of these operations can be built as a function, and our most important task at this stage in the design is to ensure that all these functions will fit together when completed. So right from the start we specify the types of these functions.

Our first function should take the original string and return a list of character lists (words), that is, a *string list list*. For example, the input string 'multiply 3268347 and 574839' would be converted to

```
[
    ["m","u","l","t","i","p","l","y"],
    ["3","2","6","8","3","4","7"],
    ["a","n","d"],
    ["5","7","4","8","3","9"]
]
```

The next function will take the *string list list* and extract from it a keyword and two numbers. The keyword could be coded in some way, so that alternatives like 'multiply' and 'times' can have the same code. The result of this function, then, would be *code * int * int*. The arithmetic function would take the code for the operation and the two integers and produce an integer result. Finally, the output function would take the integer and produce a string.

Before plunging into the implementation of these functions, let's consider more carefully the costs involved. Our first function constructs a list of character lists, only to have our second function deconstruct it again. This seems a waste of both space and time. Can we get rid of the explicit list? In other words, is there some way in which we can design a *function* to represent the list of character lists, assuming that we wish to work through it from left to right?

Some sort of successor function seems called for, that will produce the next word in the list, but this cannot be a function of the last word extracted, as there is no relationship between successive words of the sentence. It can, however, be a function of the remainder of the sentence, and this gives us a revised strategy for the extraction: convert the input string to a character list (this is a standard, cheap operation), then write a function *next_word* which takes the sentence and splits off the first word, giving two character lists in reply. The reduced character list can then be used to obtain the next word, and so on.

So 'multiply 3268347 and 574839' would be initially converted to

```
["m","u","l","t","i","p","l","y"," ",
"3","2","6","8","3","4","7"," ","a","n","d"," ",
"5","7","4","8","3","9"]
```

Then applying *next_word* to this would give

```
(
    ["m","u","l","t","i","p","l","y"],

    ["3","2","6","8","3","4","7"," ",
     "a","n","d"," ","5","7","4","8","3","9"]
)
```

where the second character list could be used to give the *next* word, and so on.

Having redesigned our first function in this way, we now have to change the type of our recognition function. We still want it to produce a code and two integers, but it cannot use the result of the first function as argument. One answer would be to make the recognition function operate on the original character list, and have it use *next_word* internally to split up the list. This design seems reasonable so we list the functions:

next_word : *string list* → *string list* * *string list*
recognize : *string list* → *code* * *int* * *int*
calculate : *code* * *int* * *int* → *int*
reply : *int* → *string*

altogether making

savant : *string* → *string*

next_word

We immediately come to the question: what is a word? Well, for our purposes a word can only be a sequence of letters such as 'add' or a sequence of digits, so our *next_word* function only has to recognize these. We could, if we wished, make it distinguish between them (changing its result type), but, if we did, it would be difficult to prevent it recognising 'add123' as two (valid) words, instead of one invalid one. So let's stick to our original plan of just recognizing sequences of letters or digits, with the corollary that mixed sequences will also be recognized as (invalid, as it happens) words.

We require a function to recognize a letter or digit:

```
fun is_letter_or_digit s =
  is_letter s orelse is_digit s
;

val is_letter_or_digit = fn : string -> bool
```

Our function *next_word* can return a pair of null lists when given a null list, and otherwise has two inductive cases:

1. If the first character of the list is a letter or digit, the extracted word will have this character prefixed to the word which *next_word* would extract from the character list minus its first character.
2. If the first character of the list is not a letter or digit, then either:
   - We haven't found a valid character yet, so this character can be ignored or
   - This character is a terminator, and the extracted word is the null list.

How do we distinguish these last two cases? One way is to make two different functions to deal with the two cases: one function to strip the initial invalid characters, the other to build up the word and deal with termination.

```
fun strip_chars []    = []
|   strip_chars (s::S) =
  if is_letter_or_digit s then s::S
  else strip_chars S
;

val strip_chars = fn : string list -> string
list

fun build_word []    = ([],[])
|   build_word (s::S) =
  if not (is_letter_or_digit s) then ([],S)
  else
    let
      val (T,U) = build_word S
    in
      (s::T, U)
    end
;

val build_word = fn : string list -> string list
* string list
```

This gives the following definition for *next_word*:

```
val next_word = build_word o strip_chars;

val next_word = fn : string list -> string list
* string list
```

A quick test:

```
next_word (explode "   abc123 def456");

val it = (["a", "b", "c", "1", "2", "3"], ["d",
"e", "f", "4", "5", "6"]) : string list * string
list
```

recognize

This function is quite pernickety — it requires the sentence to contain exactly one keyword and two integers — so it will be naturally modelled as a partial function with an exception *Recognize*. It would seem to be recursive in some way on the number of words in the sentence. The recurrence relation is not exactly obvious, though — the keyword and integers can be anywhere in the sentence, with all kinds of 'noise' words in between. So we could, in general, have the structure:

junk keyword junk first_integer junk second_integer junk

We can recognize integers but not keywords, so we need a subfunction to do this. The keyword code can be defined with a datatype statement which includes all the possible codes for arithmetic operations and a code for an unrecognized keyword:

```
datatype code =
  ADD | SUBTRACT | MULTIPLY | DIVIDE | INVALID
;

datatype code
constructor ADD : code
constructor SUBTRACT : code
constructor MULTIPLY : code
constructor DIVIDE : code
constructor INVALID : code
```

and we could write this function as:

```
val word_code =
  fn ["a","d","d"] => ADD
  |  ["s","u","b","t","r","a","c","t"] =>
     SUBTRACT
  |  ["m","u","l","t","i","p","l","y"] =>
     MULTIPLY
  |  ["t","i","m","e","s"] => MULTIPLY
  |  ["d","i","v","i","d","e"] => DIVIDE
  |  _  => INVALID
;

val word_code = fn : string list -> code
```

We're still in the dark, however, on the recurrence relation for *recognize.* Let's begin by considering a few special cases. A null sentence is obviously a no–no:

```
exception Recognize
recognize [] = raise Recognize
```

A sentence headed by a junk word will give the same result as that sentence with the junk word removed:

```
recognize S =
 let
    val (first_word,rest) = next_word S
    val code = word_code first_word
  in
    if code = INVALID
    then recognize rest
    . . .
```

Once a valid keyword has been recognized, we need to look for two integers, and the sentence is only valid if we find them. We can model this case in ML by inventing a partial function *two_ints* to look for the integers, which raises *Recognize* if it fails to find them.

```
    . . .
    else
      let
        val (int1,int2) = two_ints rest
      in
        (code,int1,int2)
        . . .
```

Now we have the problem of hunting for the two integers. Clearly, a null sentence is again a poor prospect:

```
two_ints [] = raise Recognize
```

and junk must be ignored:

```
two_ints S =
  let
    val (first_word,rest) = next_word S
  in
    if not is_digit_cl first_word
    then two_ints rest
    . . .
```

and if we find an integer, we are left with the problem of recognising the second integer:

```
    . . .
    else (cl_int first_word, one_int rest)
```

By now the pattern is becoming clear, and we can write down *one_int* fairly easily:

```
fun one_int [] = raise Recognize
|   one_int S =
  let
    val (first_word,rest) = next_word S
  in
    if not (is_digit_cl first_word)
    then one_int rest
    else cl_int first_word
  end
;

val one_int = fn : string list -> int
```

and build the other functions on top of it:

```
fun two_ints [] = raise Recognize
|   two_ints S =
  let
    val (first_word,rest) = next_word S
  in
    if not (is_digit_cl first_word)
    then two_ints rest
    else (cl_int first_word, one_int rest)
  end
;

val two_ints = fn : string list -> int * int

fun recognize [] = raise Recognize
|   recognize S =
  let
    val (first_word,rest) = next_word S
    val code = word_code first_word
  in
    if code = INVALID
    then recognize rest
    else
      let
        val (int1,int2) = two_ints rest
      in
        (code,int1,int2)
```

```
        end
    end
;

val recognize = fn : string list -> code * int *
int
```

A few tests:

```
recognize (explode "multiply 34 by 45");

val it = (MULTIPLY, 34, 45) : code * int * int

recognize (explode "junk multiply junk 56 junk
12 junk");

val it = (MULTIPLY, 56, 12) : code * int * int

recognize (explode "junk multiply junk 56
junk");

ML EXCEPTION: Recognize
```

calculate

This function is fairly straightforward after the intellectual rigours of *recognize*. It takes the keyword code and uses it to calculate the result from the two arguments. Because only the valid codes can produce a result, this will be a partial function:

```
exception Calculate
fun calculate (c,j,k) =
  case c of
    ADD => j + k
  | SUBTRACT => k - j
  | MULTIPLY => j * k
  | DIVIDE => j div k
  | _ => raise Calculate
;

exception Calculate
val calculate = fn : code * int * int -> int
```

The tests of *calculate* are totally straightforward and are omitted.

reply

Again a fairly uncomplicated function:

```
fun reply n =
  "Easy! The answer to your question is " ^
  implode(int_cl n)
;

val reply = fn : int -> string
```

savant

Now we put all the functions together to make our idiot savant. The savant will attempt to produce an answer if it can, but if it can't (a situation that leads to the raising of exception *Recognize*), it returns the string 'I don't understand your question'.

```
fun savant s =
  reply(calculate(recognize(explode s)))
  handle Recognize =>
    "I don't understand your question"
;

val savant = fn : string -> string
```

In reality of course, we would combine the previous functions with *savant* using a *local* expression, but let's forget about this stage, turn our monster loose and ask it a few questions (on simple numbers, so we can check the results!):

```
savant "O idiot, please add 1 to 1";

val it = "Easy! The answer to your question is
2" : string

savant "How are you today?";

val it = "I don't understand your question" :
string

savant "subtract the number 1 from the number
2";
```

```
val it = "Easy! The answer to your question is
1" : string

savant "multiply 100 by 3"

val it = "Easy! The answer to your question is
300" : string

savant "divide 100 by 3";

val it = "Easy! The answer to your question is
33" : string

savant "divide 3 into 100";

val it = "Easy! The answer to your question is
0" : string
```

Oh well! Back to the drawing board!

Discussion

Building our idiot savant was not a wholly frivolous exercise. It is one of the largest pieces of ML code we have produced so far, yet functional decomposition (our *divide and conquer* principle) gives it a natural structure, and ensures that the individual functions are understandable, and the way in which they relate to each other is simple and intuitive. ML's insistence on strict typing is an advantage here, since it forced us, right at the beginning, to think seriously about the **interfaces** between our functions, the way in which information is passed around from function to function. The interfaces were modified slightly when the detailed design work began, but remained in the background as the skeleton on which the rest of the design was hung.

Another point which the example brings out is the interdependence of testing and proving. All of the functions were either designed using induction (*next_word*, *recognize*, *two_int*, *one_int*) or were proved by inspection (*word_code*, *calculate*, *reply*, *savant*), and therefore reflected what we wanted to happen. But is *savant* correct? It fails to deal properly with a valid order, simply because we didn't imagine that the order would ever be given. Looking at it again, we see that negative numbers would also cause it problems

```
savant "add 1 and -1";

val it = "Easy! The answer to your question is
2" : string
```

and no doubt the reader can think of other situations where it would fail. The issue is not correctness or the lack of it, but the limited nature of our imaginations, a problem no mathematical theorem can solve. So correctness, while a necessary condition for good software, is not a sufficient one; insight, imagination, flair — all these qualities play a part too. And testing remains a vital part of the software development process, simply because of those elusive cases we never thought of.

9.7 CHAPTER SUMMARY

Central to any notion of human–computer interaction is the idea of a sequence of characters or codes. The type *character list* (*string list* in ML) models this idea, and is a convenient format for holding information within the computer. Other types can be converted to and from character lists, a process that is eased by the panoply of list-processing functions (including higher-order functions) which are available.

As a by-product of type conversion from a character list, validation functions are obtained. These are partial functions on a character list which not only detect conversion errors by raising an exception but also tailor that exception to give useful information about the kind of error detected. Validation functions can be very simple or extremely complex, but their interface remains simple, however complex their internal workings become.

Using character lists, more complex problems of text analysis can be tackled. The principle of functional composition, plus the strong typing of ML, ensures that however elaborate matters become, the interfaces between the component functions, and the functions themselves, remain simple and intuitive.

Tackling these larger problems reveals that correctness of code, while essential, is not sufficient to ensure good software. Some feedback from the results of the design process into the specification process is usually necessary to obtain robust code.

EXERCISES

1. Write a function *is_letter* which returns *true* if and only if a character is a letter.
2. Write a function upper which will convert all lower-case characters in a character list to their equivalent upper-case characters, leaving other characters untouched.
3. Write a function valid_op : string list Æ string which checks whether an input character list is a valid arithmetic operation: +, -, * or /, and returns the operation character or a Verr exception.
4. Write a function valid_day : string list Æ day which accepts a three-letter code (mon, tue, etc.) and converts it to a value of type day, giving a suitable Verr exception on invalid code.
5. Design and write a function valid_code : string list Æ int to validate a code number which must lie in the range 1000 to 9999. The function should give a suitable Verr exception on failure to convert.
6. Design and write a function valid_time : string list Æ int * int, which validates and converts any time in the format hh:mm (24-hour clock).
7. Design and write a function valid_word : string list Æ string list to validate a word. A word is defined to consist of alphabetic characters with possible embedded hyphens, apostrophes and underline characters. By 'embedded' I mean that these characters are not allowed to be the first or last character in a word, so O'Neill-Phillip's_brother is a valid word, but authors' is not.
8. Describe how you would modify the idiot savant so that it can deal with the order 'divide 3 into 100'.
9. Describe how you would modify the idiot savant to deal with orders like 'multiply –6 by –5'.

CHAPTER

TEN

INPUT AND OUTPUT

10.1 INTRODUCTION

In this chapter we move on from simple functions which fit the read–evaluate–print paradigm of interactive functional languages to more complex functions where this paradigm breaks down. Consider a complex interactive program such as a video game or an airline booking system; there is no way in which the pattern of input and output can be sensibly reduced to the input of an expression, the evaluation of its value and the output of the result.

Does this mean that functional languages are inappropriate for such applications? Not at all, but the straightforward read–evaluate–print paradigm has to be modified to cope with the increased interaction.

10.2 THE READ–EVALUATE–PRINT PARADIGM

When we type an expression for ML to evaluate, we expect the answer to appear below the expression:

```
sqrt 2.0;

val it = 1.414214 : real
```

Instead of the expression, we could have written a declaration:

```
val it = sqrt 2.0;

val it = 1.414214 : real
```

and in fact the two forms are completely equivalent. So now we realize that all our interactions with ML have been of this simple form: we have declared a named value (the default declaration being the value *it*) and ML has evaluated the declaration. Before it can evaluate the declaration, ML must read it; after it has evaluated it, it prints the answer. In this way we gain the interactive form of program development which is one of the strengths of a functional language. Now we wish to model more complex interactions, while retaining the simplicity of this interactive approach.

10.3 STREAMS

These more complex interactions can be modelled explicitly by using **streams**. A stream is a finite or infinite sequence of characters; if finite, it may or may not be terminated. There are two standard streams in ML: *std_in*, which is an input stream associated with the keyboard, and *std_out*, which is an output stream associated with the screen (Fig. 10.1). They have types

```
std_in : instream
std_out : outstream
```

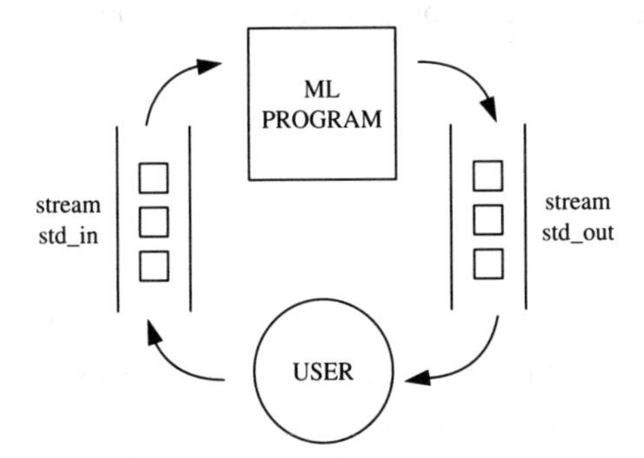

Fig. 10.1 Standard streams in ML

(*instream* and *outstream* are basic ML types). *std_in* and *std_out* act as **buffers** between the user and the ML program, allowing the user to correct mistakes, for example. In general, the characters outstanding in the input stream are delivered to the program, in order, when the user presses the RETURN key. The characters in the output stream are sent, in order, to the screen as quickly as the hardware allows.

The user can declare other, non-standard, streams, and we shall see an example of this in Chapter 12. But for the moment we consider only the case of a single user, sitting at a terminal, interacting with a single ML program.

Standard Functional Input/output

For the single-user, single-program case, it will be convenient to treat the standard streams *std_in* and *std_out* together as a pair, so we define a new type

```
datatype iostreams =
  IOSTREAMS of instream * outstream;
```

An interaction between the computer and the user will consist of the output of some information by the computer, followed (eventually) by the input of some information from the user. We can model this as a function:

$f : 'a \rightarrow iostreams \rightarrow iostreams * 'b$

This is a function which takes some value of type *'a* (the value to be output by the computer) and the current value of the streams, and returns a pair consisting of the updated value of the streams and some value of type *'b* (the value input by the user).

We can also consider two special cases: functions which input only:

$fi : iostreams \rightarrow iostreams * 'a$

and functions which output only:

$fo : 'a \rightarrow iostreams \rightarrow iostreams$

Input function *fi* takes the current value of the streams, and returns the updated streams and a value of type *'a*; output function *fo* takes a value of type *'a* and the current value of the streams, and returns the updated streams. The currying of the arguments is deliberate, and brings advantages, as we shall see later on.

Because the streams are connected to the keyboard and screen, the application of these functions is synchronized to the operation of the hardware; an input function will not return a result until the user has delivered the requisite characters (or terminated the stream in some way), and an output function will not return a result until the output stream has been updated and the characters transferred to the screen.

10.4 *get_line* AND *put_string*

We now define a standard input function and a standard output function. These functions are geared to the kind of input and output we want to do in the single-user, single-program situation. For input, we would like a function which takes a stream pair and delivers the first line which the user types (along with the updated stream-pair):

get_line : *iostreams* → *iostreams* * *string list*

A typical declaration involving this function would be

val (*v1*,*S*) = *get_line v*

(We use the letter *v* to represent the stream-pair, as it is unlikely to be confused with any other type.)

- *get_line* removes the first text line of the input stream of *v*, and returns it as a character list *S*, along with the updated streams *v1*
- The result is returned as a character list rather than a string to facilitate analysis using standard functions

For output, we simply require a way of adding a string of characters to the end of a stream. We use a string as argument rather than a character list because it is easier to implode character lists than explode string literals:

put_string : *string* → *iostreams* → *iostreams*

A typical declaration involving this function would be

val *v1* = *put_string s v*

- *put_string* appends string *s* to the output stream of *v*, and returns the updated stream pair *v1*

Implementing *get_line* and *put_string* in ML

The reader may have noticed that no ML definition has been given for these two functions. There is a good reason for this — the functions are impossible to implement in ML! Standard ML has a non-functional form of input and output. However, we would like to remain as close to the functional model of input and output as possible, so that we can use our usual proof methods, and so we use a pseudo-implementation of the functions *get_line* and *put_string*, and indicate where this pseudo-implementation is less than fully functional. The pseudo-implementation is given in Appendix 5.

Suppose we wish to obtain the user's name, consisting of surname and forename. We could write the following function to perform this interaction:

```
fun get_name v =
  let
    val v1 =
      put_string "Please enter your forename: " v
    val (v2,forename) = get_line v1
    val v3 =
      put_string "Please enter your surname:" v2
    val (v4,surname) = get_line v3
  in
    (v4, forename, surname)
  end
;

val get_name = fn : iostreams -> iostreams *
string list * string list
```

In this function the names *v*, *v1*, *v2*, *v3* and *v4* represent different values of the input/output stream-pair. But it is clear that only the latest value is used at any stage; *v* is transformed to *v1* and then forgotten, *v1* is transformed to *v2* and then forgotten, and so on. If we obey the rule of only using the latest (i.e. highest numbered) value of the iostreams at any point in the function, the ML pseudo-implementation will work. Previous values of the iostreams are lost by the implementation, and any reasoning involving these previous value will be incorrect as far as the ML implementation goes.

So long as we obey this restriction, we can use our standard proofs by induction, inspection, exhaustion and substitution to check our code, just as for functions which do not use streams.

Note that in the above example, and all those that follow, the order in which the declarations are made in the let-expression is the same as the order in which the input and output is actually performed by the implementation.

We would now like to test this function, but before we do so, we must set up the standard input/output stream-pair which the function uses as argument. There is a function called *iostreams* in Appendix 5 which will do this for us:

```
val io = iostreams std_in std_out;

val io = IOSTREAMS(-, -) : iostreams
```

By convention, the initial value of the standard/input output stream-pair is known as *io*. Now we can test *get_name*:

```
val (v,forename,surname) = get_name io;

 Please enter your forename:  fred
 Please enter your surname: bloggs

val v = IOSTREAMS(-, -) : iostreams
val forename = ["f", "r", "e", "d"] : string
list
val surname = ["b", "l", "o", "g", "g", "s"] :
string list
```

10.5 COMPOSING I/O FUNCTIONS

Often, when conversing with the user, we wish to output a whole series of strings and then input a reply. The curried form of the input/output functions makes this easy. For example, suppose we want to output a menu and ask the user to make a choice from it. If we partially apply *put_line* to a string, we obtain a function that takes a stream-pair and returns a stream pair. Any number of these partially applied functions can then be composed together.

Furthermore, function *get_line* also takes a stream-pair as argument, so it can be composed with the partially applied *put_lines* to give the final result:

```
fun get_choice v =
  let
    val (v1,choice) =
      (put_string "Select your pizza:\n\n" &
       put_string "1. Margherita\n" &
       put_string "2. Napolitana\n" &
       put_string "3. Quattra Stagione\n\n" &
       put_string "Your selection:" &
       get_line) v
  in
   (v1,cl_int choice)
  end
;

val get_choice = fn : iostreams -> iostreams *
int
```

As you can see, the forward functional composition operator is more convenient here. Of course, in this case we could have used string concatenation to save some of the composing:

```
fun get_choice v =
  let
    val (v1,choice) =
      (put_string
        ("Select your pizza:\n\n" ^
         "1. Margherita\n" ^
         "2. Napolitana\n" ^
         "3. Quattra Stagione\n\n" ^
         "Your selection:"
        ) &
       get_line) v
  in
   (v1,cl_int choice)
  end
;

val get_choice = fn : iostreams -> iostreams *
int
```

but in general, this ability to compose input/output functions can be very useful. A quick test:

```
val (v,ch) = get_choice io;
```

```
Select your pizza:

1. Margherita
2. Napolitana
3. Quattra Stagione

Your selection: 2

val v = IOSTREAMS(-, -) : iostreams
val ch = 2 : int
```

Exercise Rewrite function *get_name* using composition.

Answer

```
fun get_name v =
  let
    val (v1,F) =
      (put_string
         "Please enter your forename: " &
       get_line) v
    val (v2,S) =
      (put_string
         "Please enter your surname: " &
       get_line) v1
  in
    (v2,F,S)
  end
;

val get_name = fn : iostreams -> iostreams *
string list * string list
```

10.6 RECURSIVE INPUT/OUTPUT FUNCTIONS

I have blithely stated that recursive functions can be written, and inductive proofs performed, on functions using streams. Let's now write a function to check this statement. Suppose we wish to collect names from a user and put them into a list, the process coming to an end when the user enters 'stop'. This is a recursive function; in fact the method of differences gives us the form of the function (and the proof) — as the list increases in size, each item prefixed to the list requires a conversation with the user in which a name is requested and input. The base case (null list) corresponds to the user immediately saying 'stop'. Even in the base case, we notice, there is a conversation between the computer and the user.

The function, which we can call *list_names*, takes the current streams as argument, and returns the updated streams and the list.

```
fun list_names v =
  let
    val (v1,name) =
      (put_string "Please enter next name: " &
       get_line) v
  in
    if name = ["s","t","o","p"] then (v1,[])
    else
      let
        val (v2,name_list) = list_names v1
      in
        (v2, name :: name_list)
      end
  end
;

val list_names = fn : iostreams -> iostreams *
string list list
```

For the formal inductive proof, we need some conventions for the representation of streams. As a stream is a sequence of characters, we can represent it as a string, and for convenience we will use ML's string conventions. We are not concerned with stream termination, so all our streams will be unterminated and possibly infinite — this does not affect the proof, as we shall see. In this particular problem, we are not concerned about the value of the output stream, so we use the wild card value _ to represent it. The stream-pair v will be represented as a pair $(s,\_)$ consisting of the current input stream and output stream respectively.

Note that under these conventions we have

get_line $(u$ ^"\n" ^ $s,t) = ((s,t),$*explode* $u)$
where u is a string not containing the character "\n"
put_string $u\ (s,t) = (s,t$ ^ $u)$

Base case To prove that *list_names* ("stop\n" ^ $s,\_) = ((s,\_),[])$.
By the above definition of *get_line* and *put_string*, the result of the first *val* declaration will be

$v1 = (s,\_)$
$name = [\text{"s"},\text{"t"},\text{"o"},\text{"p"}]$

so the result of the function will be $((s,\_),[])$

Inductive case To show that, if *list_names* $(s,\_) = ((t,\_),C)$, where C is a (possibly null) list of character lists corresponding to a list of names terminated by 'stop' in s, then

list_names $(u$ ^ "\n" ^ $s,\_) = ((t,\_),$*explode* $u :: C)$

where u is a name (a string not containing a "\n" character and not equal to "stop").

Consider *list_names* $(u$ ^ "\n" ^ $s,\_)$. By the definition of *get_line* and *put_string* we have

$v1 = (s,\_)$
name = *explode u*

Because $u \neq$ "stop", we have

$v2 = (t,\_)$ by the inductive assumption
name_list = C

so the result of the function will be

$((t,\_),$*explode* $u :: C)$

QED

As you can see, the proof is quite straightforward, because each new variable introduced has a unique name. (If we wanted to re-use names, we would have to use the rule that the value of the name at any point in the code would be that given by the nearest preceding value declaration. It is easier not to re-use names!) The proof is a valuable check on our code, but we will do a test just to make sure we have not missed anything:

```
val (v,name_list) = list_names io;

Please enter next name:  matilda
Please enter next name:  doris
Please enter next name:  ernest
Please enter next name:  stop
val v = IOSTREAMS(-, -) : iostreams
val name_list = [["m", "a", "t", "i", "l", "d",
"a"], ["d", "o", "r", "i", "s"], ["e", "r", "n",
"e", "s", "t"]] : string list list
```

10.7 VALIDATED INPUT

We are now in a position to put together some of the ideas of this chapter and the previous one; in other words to write a function to perform validated input from a stream-pair. We already have our validation functions of type *string list* → *'a*; these are partial functions, remember, which raise exception *Verr* when they attempt to convert an invalid character list. The *Verr* exception has a string argument which takes the form of a handy hint as to what is wrong with the character list.

Our validated input function, which we can call *prompt*, will take a stream-pair as argument, for sure, but we can also make it very general by including the prompting string to be output, and the validation function to be applied to the user's reply, as arguments too. This gives us a function of type (*string list* → *'a*) → *string* → *iostreams* → (*iostreams* * *'a*). I have made the validation function the first argument as I can imagine the situation in which we want to partially apply *prompt* to one argument only (to make, for instance, a function to prompt for integers only). This function could then be further tailored with various prompting strings.

The prompting function will only return the user's reply if the validation function accepts it; otherwise it must intercept the *Verr* exception and repeat the prompt (including for good measure the handy hint) as many times as are necessary for the user to produce an acceptable reply. So *prompt* is a recursive function whose base case occurs when the input stream is headed by an acceptable reply. In the inductive case, the stream is headed by a sequence of unacceptable replies, followed by an acceptable one. The key point is that this sequence of replies is exactly equivalent to one acceptable reply, so a *prompt f s* on a stream headed by an unacceptable reply is equivalent to a *prompt f s* on the stream with that reply removed from it.

Having found the form of the recurrence relation, we can write the function:

```
fun prompt f s v =
  let
    val (v1,reply) = (put_string (s ^ ": ")  &
                          get_line) v
  in
    (v1,f reply)
    handle Verr t =>
      (put_string ("Reply " ^ t ^ "\n\n") &
       prompt f s) v1
  end
;

val prompt = fn : (string list -> 'a) -> string
-> iostreams -> iostreams * 'a
```

The result of the handle clause is an expression of the same type as the expression to which the handle clause is attached, so the function always returns a result of type *iostreams * 'a*.

```
val (v,number) =
  prompt valid_integer
         "Please enter a number"
         io
;

Please enter a number:  hello
Reply not an integer number

Please enter a number:  80
val v = IOSTREAMS(-, -) : iostreams
val number = 80 : int
```

This function is significant in two ways: it provides a user-friendly interface, which allows for the fact that the user is going to make mistakes; and it eases the task of proof in functions which use it. Because *prompt* embodies the recurrence relation

$$prompt\ f\ s\ (invalid\_reply\ \wedge\ t,\_)\ =\ prompt\ f\ s\ (t,\_)$$

(where we use our standard technique of modelling a stream by a string) we can simply ignore the possibility of the user making invalid replies in proofs of functions which use *prompt*.

10.8 MENUS

One very common method of interaction with a user is the **menu**. The user is presented with a fixed set of alternatives to choose from. Because the choice is limited, the validation required is minimal, indeed, for a WIMP interface (that is, an interface relying on Windows, Icons, Menus and Pointing) no validation is required at all, giving a very direct and user-friendly interface.

The system proposed here is not as sophisticated as a WIMP interface. It does have the advantage of working on a wide range of computer terminals.

Here is an example of a menu screen, taken from the library case study of Chapter 11:

```
                    Library System Main Menu

        0. Exit

        1. Issue library card
        2. Add new book to library
        3. Lend book
        4. Return book
        5. Renew book
        6. Reserve book
        7. Cancel reservation
        8. Lend from reservation pile
        9. Display data

       Please enter your selection:
```

Entering the appropriate number causes the function corresponding to the prompt to be activated, and the menu to be redisplayed.

Menus can form a hierarchy — for example, if the user selects option 9 of the above menu, a submenu is displayed:

```
                          Display data

        0. Exit

        1. Display users
        2. Display books
        3. Display loans
        4. Display reservations
        5. Display pile of reserved books

       Please enter your selection:
```

The user can select options (functions) from this menu, or, by selecting zero, can return to the previous menu.
In general then, we can have a complete tree of menus, with the selected functions forming the leaves of the tree, as shown in Fig. 10.2. This is a recursive structure, so it should be possible to use a recursive function to implement it. Let's think for a moment about the type of this function. What a menu-based system allows us to do is apply selected functions, one after the other, to the data we are interested in. As any

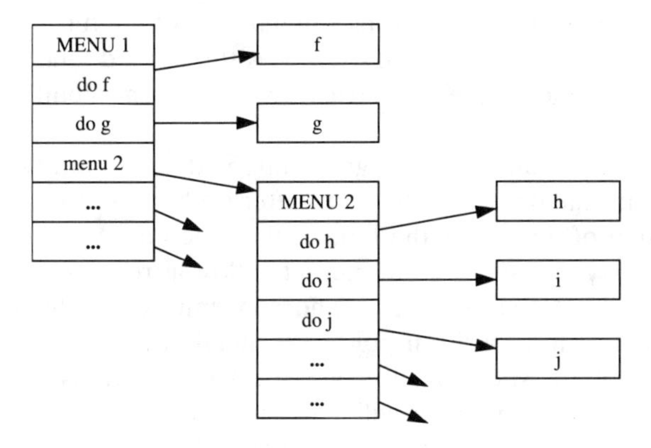

Fig. 10.2 A tree of menus

function can be applied to the result of any function (including itself) the functions must all have the same type for argument and result, in other words they must all be of type $'a \to 'a$. The overall selection function is a (possibly null) composition of functions of type $'a \to 'a$, so it must also be of this type. This means that another instance of the selection function can itself be an option, as in Fig. 10.2.

But what about the title of the menu, and the details of its options? Surely these must also be arguments of the selection function? The paradox can be resolved by making the selection function a curried function, and then partially applying it to obtain a function of type $'a \to 'a$.

There is one more aspect to the argument. The user is communicating with the selection function and the other functions of the system by means of the standard stream-pair. To make these interactions coherent, the same stream-pair must be used by all functions, and must be taken as argument and passed as result. This gives a type for the leaf functions of *iostreams* * *'a* → *iostreams* * *'a*. In addition, the selection function needs:

- A title for the menu, of type *string*
- A list of option strings, of type *string list*
- A list of option functions, of type (*iostreams* * *'a* → *iostreams* * *'a*) *list*

The selection function therefore has a type of *string* → *string list* → (*iostreams* * *'a* → *iostreams* * *'a*) *list* → *iostreams* * *'a* → *iostreams* * *'a*.

The *select* Function

We are now going to implement the menu selection function, which we will call *select*. It is a recursive function, in the sense that its normal operation is repetitive — display menu, apply function, display menu, and so on. It is also potentially recursive in another sense, in that one of the functions it applies may be a specialized version of itself. It is this second kind of recursion that allows us to model the tree-structure of a typical menu system.

One of the nice features of functional programming is that we don't have to worry about the internal details of functions after we have written them. The only information of interest is the type of the function. As we have carefully designed the type system of our menu to allow a tree structure, we can now forget about that structure, and concentrate on the process of selection. Once we have written our *select* function correctly, we can guarantee that a suitably specialized version of it will serve as a valid menu option, simply because the types will match up correctly.

To fix ideas, let's look at a typical declaration involving the function:

```
val (v1,a1) = select t S F (v,a)
```

Here, *t* is the title string, *S* is the list of options, *F* is the corresponding list of functions, *v* is the initial value of the stream-pair, and *a* is the initial value of the data type the menu system is being applied to (such as the library system in the example above).

Base case This occurs when the user selects the zero option. In this case, *a1* = *a*, and *v1* is simply the latest version of the stream-pair. Formally,

select t S F ((“0” ^ “\n” ^ *s*, _), *a*) = ((*s*, _), *a*)

Inductive case This occurs when the user selects some other option. In this case, the selected function from the list *F* is applied to *v* and *a*, giving values *v1* and *a1*. The *select* function is then applied to these updated values, allowing other option functions to be applied as many times as required.

Formally, if F_i is the ith entry in the function list, and “*i*” is the string corresponding to the integer *i*, then

if *select t S F* (F_i ((*s*,_),*a*)) = ((*s1*,_),*a1*)
then *select t S F* (“*i*” ^ “\n” ^ *s*,_),*a*) = ((*s1*,_),*a1*)

This recurrence relation determines the structure of the function and the proof.

In both cases, the function must display a menu and accept the user's reply. We assume that the following functions have been already implemented:

- *put_title* : *string* → *iostreams* → *iostreams*, which outputs a title, centred on the screen
- *menu_string* : *string list* → *int* → *string*, which converts the list of options into a menu string, the options being labelled with integers starting at the value of the second argument
- *valid_range* : *int* → *int* → *string list* → *int*, which validates that a character list is an integer in the range given by the first two arguments
- *nth* : *int* → *'a list* → *'a*, which returns the nth item of the list (counting from one), the first argument giving the value of n.

For the base case of the function we have

```
fun select t S F (v,a) =
  let
    val (v1,i) =
      (put_title t &
       put_string ("0. Exit\n\n" ^
                   (menu_string S 1) ^
                   "\n") &
       prompt (valid_range 0 (length F))
              "Please enter your selection"
      ) v
  in
    if i = 0 then (v1,a)
    ...
```

If the user's reply is not zero (it is bound to be in range) then the ith function is selected from the list, applied to *v1* and *a*, and *select* is re-applied to the result of this function.

```
    ...
    else
      let
        val Fi = nth i F
      in
        select t S F (Fi (v1,a))
      end
  end
;

val select = fn : string -> string list ->
(iostreams * 'a -> iostreams * 'a) list ->
iostreams * 'a -> iostreams * 'a
```

A test, just to be on the safe side:

```
val (v,newval) = select "test menu"
                        ["increment"]
                        [fn (v,i) => (v,i+1)]
                        (io,46)
;
                  test menu
_____________________________

0. Exit

1. increment

Please enter your selection:  1


                  test menu
_____________________________

0. Exit

1. increment

Please enter your selection:  0
val v = IOSTREAMS(-, -) : iostreams
val newval = 47 : int
```

A real system would hopefully do something less trivial, but the principle of operation has been demonstrated.

I hope that these examples have convinced the reader that using a functional language for interactive input and output is not only realistic but, thanks to the use of higher-order functions and currying, can also give quite elegant code.

10.9 CHAPTER SUMMARY

> Complex interactive systems require something more than the read–evaluate–print paradigm. ML provides streams to enable these systems to be implemented. A simple (pseudo) functional input/output system has been demonstrated, based on the concept of a stream-pair.
>
> With care, interactive programs can be written in such a way that proofs can be performed, using the full panoply of proof techniques, including induction.
>
> The use of higher-order functions and currying eases the implementation of interactive functions, as was demonstrated by a function to prompt for validated input, and a menu-selection function.

EXERCISES

In these exercises the functions *get_line* and *put_string* from Appendix 5 and the validation functions from Appendix 4 should be used.

1. Modify function get_name so that having asked the user his first name, and received the reply, say 'Richard', asks 'Richard who?' to obtain the second name. The functional spec should remain the same.
2. Write expressions involving the function prompt to obtain the following information:
   (a) User's forename (all letters)
   (b) User's date of birth
   (c) User's local telephone number
   (d) User's height in metres
   (e) Direction user's garden faces (N, S, E, W).
3. Design and write the functions *put_title* and *menu_string*, which were mentioned above as sub-functions of *select*. Their specifications are:
   - *put_title* : *string* → *iostreams* → *iostreams*, outputs a title, centred on the screen. Assume that the width of the screen is given by *screen_width* : *int*.
   - *menu_string* : *string list* → *int* → *string*, converts a list of options into a menu string, the options being labelled with integers starting at the value of the second argument.

4. Write a declaration for a tailored version of the select function that will produce the menu of the library system given above, reduced so that
   - The only options on the main menu are 'Issue library card' and 'Display data'
   - The only options on the display menu are 'Display users' and 'Display books'

   Give meaningful names to the various functions involved.
5. Given f S = a, where f is a validation function, S is a string list and a is of arbitrary type, prove that

prompt f t v = (*v1*,*a*)

if the input stream of *v* is headed by *implode S* ^ "\n".

CHAPTER
ELEVEN
CASE STUDIES

11.1 INTRODUCTION

In this chapter we look at three related case studies which build on the achievements of previous chapters. The end result will be a simple prototype of a library system which, nevertheless, is fairly sophisticated in its interactions with the user.

To deal with these larger systems we need to step outside the confines of the ML Core language we have used so far. This is because we shall be building the system as a set of components, or modules, each module consisting of a collection of functions. To guide us in the design of a module, we shall return to the notion of a data type which was first discussed in Chapter 7. This, you may recall, consisted of a set of functions which constructed, analysed and modified values of a given type. (We studied stacks, queues and lists as typical examples of datatypes.)

If we want to build a system as a collection of modules, then it would be very convenient if we had some named construct in our language which corresponded to our notion of a module. The total system could then be constructed by reference to module names, without mentioning individual function names. (Of course, *running* the system would involve reference to the function name which represents the total system.)

In passing, it is worth pointing out that none of the constructs we have met so far will suffice to represent a module. This is because a module can contain both type declarations and (function) value declarations, and there is no way in which types and values can be combined in any construct we have met so far.

ML provides the **structure** construct to allow us to build modules. An example of a structure declaration for the type *'a stack* of Chapter 7 would be

```
structure Stack = struct

  infixr 5 PUSH;
  datatype 'a stack = EMPTY_STACK
                    | op PUSH of 'a * 'a stack
  ;

  exception Pop
  fun pop (a PUSH A)   = (a,A)
  |   pop EMPTY_STACK = raise Pop
  ;

  fun is_empty_stack EMPTY_STACK = true
  |   is_empty_stack _           = false
  ;

end (* Stack *);

structure Stack : sig
  datatype 'a stack
  constructor EMPTY_STACK : 'a stack
  constructor PUSH : 'a * 'a stack -> 'a stack *
  exception Pop
  val is_empty_stack : 'a stack -> bool
  val pop : 'a stack -> 'a * 'a stack
end
```

Now if we wish to use the components of the *Stack* structure anywhere else in our program, we can say

```
open Stack;
```

and the contents of the structure are available to us without specific declaration.

*Note that the infix status of PUSH is local to the structure. This is a feature of Standard ML. Fortunately, there are ways of circumventing this restriction, which we will see later.

The Modules portion of ML contains other constructs which are used to achieve the kind of flexibility and power at the module level which functions give at the level of ordinary values. We shall not examine these other constructs here; the reader is referred to the Bibliography, especially Paulson's book, for further details.

(**Note** for readers with an implementation of the ML Core only: your compiler will not recognize the above structure declaration, but most implementations of ML Core have an extension to perform the simple inclusion of code we use in this chapter.)

11.2 SIMPLE RELATIONAL DATABASE

Before plunging into the design of our database, it will be as well to review some relational database ideas. As the name implies, a relational database is based on the mathematical idea of a relation. We have met relations already in Chapter 2, where we discovered that a function is a many-to-one relation. All the relations in Chapter 2 were binary relations between an argument type and a result type, but it is perfectly possible to have relations between three or more types (**n-ary relations**). For example, a book in the library might have **attributes** of the following types: *ISBN*, *title*, *author*. This could be represented as the relation shown in Fig. 11.1.

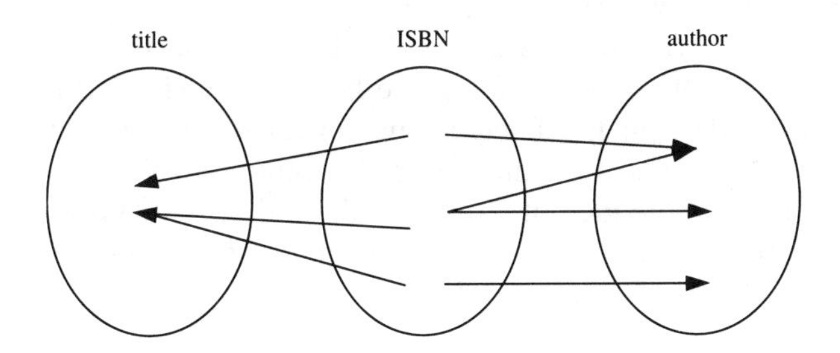

Fig. 11.1 The relation *BOOK*

This **ternary relation** can be considered to be two binary relations. Are these relations many-to-one? Let's consider the relation between titles and ISBNs. It is possible for two books with different ISBNs to have the same title, but impossible for two books with different titles to have the same ISBN, so the relation is many-to-one from *ISBN* to *title*. In other words, the title of a book is **functionally dependent** on its ISBN.

In the case of ISBNs and authors, though, the situation is different. It is possible for two books with different ISBNs to have the same author, but it is also possible for two or more authors to write a single book with a single ISBN. So here we have a many-to-many relation, and it is not possible to obtain a functional dependency between authors and ISBNs.

So, supposing Bloggs and Cloggs jointly author a book called *Fundamentals of ML*, and this book has the ISBN 0-123-45678-9, while Bloggs, branching out on his own, pens a smaller opus called *A Short Guide to ML* (ISBN 9-876-54321-0), then we have the following entries in our relational database:

title	*ISBN*	*author*
Fundamentals of ML	0-123-45678-9	Bloggs
Fundamentals of ML	0-123-45678-9	Cloggs
A short guide to ML	9-876-54321-0	Bloggs

Each line in the table is called a **tuple**, and the entire table represents a **relation**. Every finite relation, of whatever kind (one-to-many, many-to-one, many-to-many) can be represented in this way, as a set of tuples. The presence of a tuple in the set means that the corresponding relation holds between the constituent values of the tuple.

We can see that there is redundancy in this database, in that the title and ISBN for the book by Bloggs and Cloggs are held twice. This is the price we pay for storing data in the fixed format of an n-ary relation. Techniques have been developed for minimizing this redundancy (interestingly enough, they involve the idea of a functional relationship) but we shall not concern ourselves with them here. The interested reader is referred to any good book on relational databases.

Representing a Relation in ML

At first blush, it seems that the obvious way to represent a relation in ML is as a list of tuples. We can imagine a declaration

```
datatype book_tuple =
   BOOK_TUPLE of title * isbn * author
;

datatype book = BOOK of book_tuple list;
```

The problem with this approach comes when we try to design functions to operate on the tuple. We would need a new set of functions for each relation we invented. If we model the relation using only built-in types the situation is hardly any better:

```
val book_relation =
  [
   ("Fundamentals of ML",0123456789,"Bloggs"),
   ("Fundamentals of ML",0123456789,"Cloggs"),
   ("A short guide to ML",9876543210,"Bloggs")
  ]
;

val book_relation = [("Fundamentals of ML",
123456789, "Bloggs"), ("Fundamentals of ML",
123456789, "Cloggs"), ("A short guide to ML",
9876543210, "Bloggs")] : (string * int * string)
list
```

Now our functions would only work on the type (*string* * *int* * *string*) *list*, and it is unlikely that all the relations we could ever invent would be of this type.

We are forced to take an extremely general view of what a relation is. In general, the tuples can be of any length from one upwards (yes, unary relations are possible! They represent a partial identity relation), and their constituent values can be of any type. Furthermore the relation itself can contain any number of tuples from zero upwards. So we have to model:

- The constituent values by a type that can be converted to any other type
- A tuple as a type that can be of indefinite length
- A relation as a type that can be of indefinite length, including zero.

The only ML types that fill the bill as far as the first condition is concerned are *string* and *character list*; we choose *string* as being the more compact representation. For the second and third conditions, we require a dynamic type, and the obvious choice from the types we have met is *list*. So we end up with the type *string list list* to model a relation, with the proviso that a tuple must have a length of at least one.

Of course, if we were building a full-scale implementation of a library system, we would put in strict type-checking:

```
datatype tuple = TUPLE of string list;

datatype relation = RELATION of tuple list;
```

and so on, but for our simple prototype, the built-in types will suffice. One useful feature that ML provides in this situation is **type binding**. This is similar to the synthetic use of *let* mentioned in Chapter 7, in that a complex expression is replaced by a single identifier, but this time the expression is a type expression, and the identifier is a type name. We could declare

```
type tuple = string list;

type relation = tuple list;

eqtype tuple = string list
eqtype relation = tuple list
```

(*eqtype* is short for *equality type*) and the names on the left-hand side will be treated as equivalent to the expressions on the right-hand side hereafter. This is a long way short of strict typing, as ML cannot distinguish between the types *relation* and *string list list*, so the built-in type is not hidden from the user, and there is no obligation to use the declared form; but it does provide more meaningful type names, and can save space. So we shall use these names in what follows.

The functions we define in this chapter will allow tuples to be added and deleted, and provide a variety of searching and filtering operations on relations. Functions to save and load the database from secondary storage (treated as stacks of relations) are given in Appendix 6.

Adding a Tuple

Bearing in mind that we are representing our database as sets of tuples, we can invent useful operations on these sets. First, we need to be able to add a tuple to a set (in our library example, to add a new book to the library, for instance). This looks like an easy function:

```
fun add_tuple (S:tuple) (SS:relation) = S :: SS;

val add_tuple = fn : tuple -> relation -> tuple
list
```

but we hit a snag with duplicate tuples. Remember that in mathematics a set is a very simple construct; all one can say is that a certain item is or is not a member of a given set. A set has no ordering among its members, and no duplicates (or rather, duplicates are irrelevant). We are using a set to represent a relation precisely because it has these properties; all we require is that a certain tuple is or is not a member of a given relation.

This leads us to a dilemma: suppose we wish to represent the relation we tabulated above:

title	*ISBN*	*author*
Fundamentals of ML	0-123-45678-9	Bloggs
Fundamentals of ML	0-123-45678-9	Cloggs
A short guide to ML.	9-876-54321-0	Bloggs

Under our approved model, both the following representations would be acceptable:

[
['*Fundamentals of ML*', '0-123-45678-9', 'Bloggs'],
['*Fundamentals of ML*', '0-123-45678-9', 'Cloggs'],
['*A Short Guide to ML*', '9-876-54321-0', 'Bloggs']
]

[
['*Fundamentals of ML*', '0-123-45678-9', 'Bloggs'],
['*Fundamentals of ML*', '0-123-45678-9', 'Bloggs'],
['*Fundamentals of ML*', '0-123-45678-9', 'Cloggs'],
['*A Short Guide to ML*', '9-876-54321-0', 'Bloggs']
]

and indeed the first tuple, or any tuple could be duplicated any number of times without affecting the properties of the relation. If we wish to find whether a tuple is present in a relation, we can search for the first such tuple in the list, and ignore the rest. What about deleting a tuple from a relation? Well, all we would have to do would be to delete all the duplicate tuples from the list (a standard filtering operation). It really looks as if there is no advantage to avoiding duplicate tuples.

However, if we wish to find the *number* of tuples in a relation, this will be much more difficult if we allow duplicates in our list. On this rather flimsy pretext we avoid putting duplicates into our list. Adding a tuple now becomes a recursively defined operation. The base case occurs when the relation is empty, and for the inductive case, adding a tuple to a non-empty relation *T*::*TT*, we have two situations:

- The tuple matches *T*. In this case the original relation is returned as the result
- The tuple does not match *T*. It may match other tuples in the relation, but in any case the result of adding it will be the result of adding it to *TT*, with *T* prefixed.

```
fun add_tuple (S:tuple) ([]:relation) =
  ([S]:relation)
|   add_tuple S (T::TT) =
  if S = T then (T::TT)
            else T::add_tuple S TT
;
```

```
val add_tuple = fn : tuple -> relation -> relation
```

This definition is a little loose as it will allow tuples of different lengths (including zero length) to be added to the same relation. We can tighten it up by raising an exception if the user attempts to add a zero-length tuple or a tuple of incorrect length:

```
exception Add_tuple
fun add_tuple ([]:tuple) _ = raise Add_tuple
|   add_tuple S ([]:relation) = ([S]:relation)
|   add_tuple S (T::TT) =
  if length S <> length T then raise Add_tuple
  else if S = T then (T::TT)
  else T::add_tuple S TT
;
```

```
exception Add_tuple
val add_tuple = fn : tuple -> relation -> relation
```

```
val R = add_tuple [] [];
```

```
ML EXCEPTION: Add_tuple
```

```
val R = add_tuple ["a","b"] [["d","e","f"]];
```

```
ML EXCEPTION: Add_tuple
```

```
val R = add_tuple ["a","b","c"] [["d","e","f"]];

val R = [["d", "e", "f"], ["a", "b", "c"]] :
relation

val R2 = add_tuple ["a","b","c"] R;

val R2 = [["d", "e", "f"], ["a", "b", "c"]] :
relation
```

Adding a tuple is now quite a lengthy operation, taking an average time proportional to the number of tuples in the relation, but as we are only building a small-scale prototype, this overhead is acceptable.

Deleting a Tuple

As a relation is a set of tuples, deleting a tuple is best modelled as a set difference operation. The difference of two sets $A - B$ is the set consisting of all the members of A which are not in B. We can take the result of deleting the tuple t from relation R to be $R - \{t\}$; in other words, all the members of R which are not t. This makes *delete_tuple* a straightforward filtering operation:

```
fun delete_tuple (S:tuple) (TT:relation) =
  filter (fn T => T <> S) TT : relation
;

val delete_tuple = fn : tuple -> relation ->
relation

val R3 = delete_tuple ["a","b","c"] R;

val R3 = [["d", "e", "f"]] : relation

val R4 = delete_tuple ["a","b","d"] R;

val R4 = [["d", "e", "f"], ["a", "b", "c"]] :
relation
```

The result of the *delete_tuple* operation is guaranteed not to contain the deleted tuple.

Detecting the Presence of a Tuple in a Relation

This is our normal membership test for lists, as defined in Chapter 8:

```
fun is_member_relation (SS:relation) =
  is_item (SS:relation)
;

val is_member_relation = fn : relation -> tuple ->
bool
```

I have used one explicit argument so that the type constraint can be expressed; ML obliges by filling in the type of the other argument and the result.

Filtering Tuples on an Attribute Value

Suppose we wish to find all the books written by a certain author, or all the loans taken out by a certain borrower; we need to filter a relation on an attribute value. We can number the attributes in a tuple from one, and then use our *nth* function to retrieve that attribute, as the tuple is represented as a list.

We invent a function *filter_tuple* such that

filter_tuple n s SS = TT

where *n* is the position of the attribute
s is the value being matched
SS is the relation being searched
TT is the result of the search

This is a straightforward specialization of our *filter* function:

```
fun filter_tuple n s (SS:relation) =
  filter (fn S => nth n S = s) SS : relation
;

val filter_tuple = fn : int -> string -> relation
-> relation
```

One of the pleasant features of this function is that we can perform multiple filtering; for example, if we add an attribute *publisher* to our *book* relation:

title	*ISBN*	*author*	*publisher*
Fundamentals of ML	0-123-45678-9	Bloggs	Triggs & Trogs
Fundamentals of ML	0-123-45678-9	Cloggs	Triggs & Trogs
A Short Guide to ML	9-876-54321-0	Bloggs	Klopf Verlag

we can easily find all the books written by Bloggs and published by Klopf Verlag:

```
val selected_books =
  filter_tuple 4
                "Klopf Verlag"
                (filter_tuple 3 "Bloggs" books)
;

 val selected_books = [["A short guide to ML", "9-876-
54321-0", "Bloggs", "Klopf Verlag"]] : relation
```

Finding a Tuple with a Given Attribute

If we know we are looking for only one tuple with a given attribute, we can save time by searching rather than filtering:

```
exception Find_tuple
fun find_tuple n s (SS:relation) =
  find_first (fn S => nth n S = s) SS : tuple
  handle Find_first => raise Find_tuple
;

exception Find_tuple
val find_tuple = fn : int -> string -> relation ->
tuple

val selected_book = find_tuple 3 "Bloggs" books;

val selected_book = ["Fundamentals of ML", "0-123-45678-
9", "Bloggs", "Triggs&Trogs"] : tuple
```

11.3 FORM MANAGEMENT SYSTEM

Having designed the 'back end' of our library system, where the data is stored, it is time to consider the 'front end', the interface with the user.

We have already designed, in Chapter 10, functions for prompting the user for input, and for menu selection. One requirement we haven't considered is the input of a set of related data, such as all the data concerned with a book or a borrower. In a manual system, such data would be collected on a form, and so the equivalent computer facility is called a **form management system**.

Our form management system will be very straightforward; it will consist of one function which takes a list of requirements as argument and returns a list of data values as result. The complications will appear when we start to consider the user interface. Once we have written this function, however, we need never worry about these user interface problems again, as our solutions will be encapsulated in the function. We need only supply a list of requirements (and an input-output stream-pair) and wait for the function to deliver the results.

The Function *get_tuple*

We first consider the type of our form management function. Each data item requested will need a validation function and a prompting string so the first argument type will be ((*string list* → *'a*) * *string*) *list*. The last argument will be a stream-pair of type *iostreams*. The function will return the updated streams and a list of results; by the rules of ML, all these results will have to be of the same type. What should this type be?

Now we can re-use some of the design of our relational database, and make the type of the returned values to be *string*. Then the form management function will produce a result (a string list) which can be stored directly in our relational database. Since a great many of the lists of data values the function returns will in fact be the values of some tuple in a relation such as BOOK, this seems to be a sensible move. We can call our function *get_tuple* to emphasize the connection.

This design idea has a repercussion on the type of the validation functions we use as arguments to *get_tuple*. They must produce a result of type *string* to ensure that the overall result is a string list. Fortunately, it is easy to convert any validation function to a function of this type, since all it needs to do is *implode* its argument (if valid, of course). As a convention, we shall form the name of any validation function that produces a result of type string by appending *_st* to the name of the normal validation function. So the string version of the *valid_cl* function becomes *valid_cl_st*, and so on.

We can even write a function, *string_it*, that will convert a normal validation function to one which returns a string. The result of *string_it f*, where *f* is some validation function, will be a function that is partial on exactly the same domain as *f*, but which always returns the imploded argument if it returns anything at all. In other words, *string_it f S* attempts to apply *f* to *S*, giving an exception if this is not possible, then only if a result is forthcoming does it substitute the imploded argument for that result:

```
fun string_it f S =
  let
    val result = f S
  in
    implode S
  end
;
```

```
val string_it = fn : (string list -> 'a) -> string
list -> string
```

```
val valid_cl_st = string_it valid_cl;
```

```
val valid_cl_st = fn : string list -> string
```

Now the final type specification for our form management function is

get_tuple : ((*string list* → *string*) * *string*) *list* → *iostreams*
→ *iostreams* * *string list*

Design of *get_tuple*

We already have a prompting function which we are anxious to re-use, and we can see that *get_tuple* is a kind of mapping of *prompt* over the ((*string list* → *string*) * *string*) *list*. The only complication is that the iostreams must be passed on at each stage of the mapping process. We can write a special version of the mapping function, called *map _io*, which will carry the streams through. For a null list, this function will simply return the streams it is given, but for a non-null list, the streams will be modified systematically by the mapping of each item. The key idea is that the value of the streams *after* the first item on the list has been mapped will be the value of the streams *before* the rest of the list is mapped:

```
fun map_io f []      v = (v,[])
|   map_io f (a::A) v =
  let
    val (v1,b)  = f a v
    val (v2,B) = map_io f A v1
  in
    (v2,b::B)
  end
;
val map_io = fn : ('a -> 'b -> 'b * 'c) -> 'a list
-> 'b -> 'b * 'c list
```

Now it is a straightforward matter to define *get_tuple* using *prompt* and *map_io*. The only complication involves changing the type of *prompt* from

$(string\ list \rightarrow 'a) \rightarrow string \rightarrow iostreams \rightarrow iostreams * 'a$

to

$(string\ list \rightarrow string) * string \rightarrow iostreams \rightarrow iostreams * string$

in other words, uncurrying the first two arguments, and putting a type constraint on the validation function.

```
fun get_tuple FS v =
  let
    fun prompt' (f:string list -> string,s) =
      prompt f s
  in
    map_io prompt' FS v
  end
;

val get_tuple = fn : ((string list -> string) *
string) list -> iostreams -> iostreams * string
list
```

This is a fine and elegant function but in testing it shows an obvious deficiency:

```
val tuple = get_tuple [(valid_cl_st,"name"),
                       (valid_integer_st,"age")]
                      io
;

name:  Jhon Simth
 age:  355

val tuple = (IOSTREAMS(-, -), ["Jhon Simth",
"355"]) : iostreams * string list
```

The function allows 'typos' through without giving the user a chance to correct them. This seems a little user-unfriendly and so we would like to modify the internal workings of the function to give the user a chance to check the data that he or she has entered, and to modify it as required.

Of course, the way in which the data is entered is purely a matter between the user and the function and is of as little consequence to the rest of the system as, for example, the decision to implement *double* by addition or multiplication. This is the advantage we gain by packaging the user interface inside a function.

Improved Version of *get_tuple*

The new version of *get_tuple* is going to be more complex than the old. But if we use our principle of functional decomposition, we can deal with the difficulties one at a time.

Suppose the user has entered the slightly garbled reply given above in the test of *get_tuple*. A mechanism has to be provided to allow the user to amend one or more of the data items. To do this, the user has to be able to identify which item or items to amend. In a mouse and pointer system this would be easy, but we have decided to limit ourselves to a standard alphanumeric terminal. One way in which the appropriate item could be identified is by number, as in a menu system. We have already implemented the *select* function in Chapter 10, so this strategy opens up the possibility of re-using many of the ideas and functions developed there.

The general strategy for *get_tuple* can now be set out as follows:

1. Ask the user for the required data items, numbering each question from 1.
2. Ask the user whether the replies are satisfactory, and if not, which item is in error.
3. Re-prompt for the selected item.
4. Display all the numbered data items, including the amended one.
5. Ask the user again whether the replies are satisfactory, and so on.

This strategy gives the user plenty of opportunity to amend the data. It also is clearly recursive in form, with the base case corresponding to the user giving the 'all clear'.

prompt_list

Let us consider the component subfunctions of this function, starting with the numbered asking of questions. This function, which we can call *prompt_list*, will bear a strong similarity to our original *get_tuple*, with the addition of a line number at the start of each prompt. This small change, however, means that we cannot so easily characterize it as a mapping operation, so an explicit version is required. We provide an extra argument *n* to give the line number, and note that the result of *prompt_list n ((f,s)::FS) v* will consist of the result of prompting for *s* using *n* as the line number and *f* as the validation function, prefixed to *prompt_list* (*n*+1) *FS v1*, where *v1* is the state of the iostreams after the initial prompt has been done. The base case (output of a blank line) will occur when *FS* is the null list. In ML:

```
fun prompt_list n [] v = (put_string "\n" v,[])
|   prompt_list n
                ((f:string list -> string,s)::FS)
                v =
  let
    val (v1,t) =
      prompt f (int_string n ^ ". " ^ s) v
    val (v2,T) = prompt_list (n+1) FS v1
  in
    (v2,t::T)
  end
;

val prompt_list = fn : int -> ((string list ->
string) * string) list -> iostreams -> iostreams *
string list
```

The function *int_string*, used in the above declaration, is a modification of the standard function *int_cl* given in Appendix 3:

```
val int_string = implode o int_cl;

val int_string = fn : int -> string
```

Now we can check the *prompt_list* function:

```
val tuple = prompt_list
              1
              [(valid_cl_st,"name"),
               (valid_integer_st,"age")]
              io
;
1. name:  Jhon Simth
2. age:  355

val tuple = (IOSTREAMS(-, -), ["Jhon Simth",
"355"]) : iostreams * string list
```

valid_choice

The next function to be designed is the one to check whether the items are satisfactory, and if not, which one is in error. This is a validation function. A typical prompt will be 'OK or number to amend' and the user will reply OK or give the number. Because the number of items in the menu is variable, it will have to be given as an argument of the function. The result of the function will be an integer, with 0 coding for the OK reply:

```
fun valid_choice n ["O","K"] = 0
|   valid_choice n S        =
  valid_range 1 n S handle Verr _ =>
    raise Verr ("not OK or a valid choice in" ^
                "range 1 . . . " ^ int_string n)
;

val valid_choice = fn : int -> string list -> int

valid_choice 5 ["O","K"];

val it = 0 : int

valid_choice 5 ["5"];

val it = 5 : int

valid_choice 5 ["6"];

ML EXCEPTION: Verr("not OK or a valid choice in
range 1 . . . 5")
```

items_string

Now we implement the function to display the amended items, and here we have a piece of luck — we already have a function *menu_string*, which converts a list of strings into a numbered menu string. *items_string* is just a specialization of this function, taking the list of validation functions and prompt strings *FS*, and the data item list *T* to form a new list of prompts plus values, *ST*:

```
fun items_string FS T =
  let
    val ST =
      combine (fn ((f,s),t) => s ^ ": " ^ t)
              (FS,T)
  in
    menu_string ST 1
  end
;

val items_string = fn : ('a * string) list ->
string list -> string

items_string [(valid_cl_st,"name"),
              (valid_integer_st,"age")]
             ["Jhon Simth", "355"]
;

val it = "1. name: Jhon Simth\n2. age: 355\n" :
string
```

amend_items

We have now manufactured the components of our new function, and it is time to start fitting them together. If we look again at our strategy for the new *get_tuple*, we can write in the component functions we will use:

1. Ask the user for the required data items, numbering each question from 1 — *prompt_list.*
2. Ask the user whether the replies are satisfactory, and if not, which item is in error — *prompt* with *valid_choice.*
3. Re-prompt for the selected item — find appropriate prompt string and validation function using *nth*, then use *prompt.*
4. Display all the numbered data items, including the amended one — *items_string* to make the string, *put_string* to display it.
5. Ask the user again whether the replies are satisfactory, and so on — er . . . .

It is clear that we need one more function which recursively performs steps 2, 3 and 4. This function will take the prompting list and the list of item values produced by *prompt_list* as arguments (and the iostreams of course), and will produce a final validated version of the item list (plus the updated streams). In the base case, the user accepts the original list.

```
fun amend_items FS T v =
  let
    val (v1,choice) =
      prompt (valid_choice (length T))
             "OK or number to amend"
             v
  in
    if choice = 0 then (v1,T)
    else
      let
        val (f,s) = nth choice FS
        val (v2,new_item) = prompt f s v1
        val U = new_nth choice new_item T
        val v3 =
          put_string
            ("\n" ^ items_string FS U ^ "\n") v2
      in
        amend_items FS U v3
      end
  end
;

val amend_items = fn : ((string list -> string) *
string) list -> string list -> iostreams ->
iostreams * string list

amend_items [(valid_cl_st,"name"),
             (valid_integer_st,"age")]
            ["Jhon Simth", "355"]
            io
;

OK or number to amend:  3
Reply not OK or a valid choice in range 1 . . .2
OK or number to amend:  1
name:  John Smith

1. name: John Smith
2. age: 355

OK or number to amend:  2
age:  35
```

```
1. name: John Smith
2. age: 35

OK or number to amend:  OK
val it = (IOSTREAMS(-, -), ["John Smith", "35"]) :
iostreams * string list
```

Final Version of *get_tuple*

Now we preface the *amend_items* function by *prompt_list* to obtain the final version of *get_tuple*:

```
fun get_tuple FS v =
  let
    val (v1,T) = prompt_list 1 FS v
  in
    amend_items FS T v1
  end
;

val get_tuple = fn : ((string list -> string) *
string) list -> iostreams -> iostreams * string
list

val T = get_tuple [(valid_cl_st,"name"),
                    (valid_integer_st,"age")]
                   io
;

1. name:  Jhon Simth
2. age:  355

OK or number to amend:  1
name:  John Smith

1. name: John Smith
2. age: 355

OK or number to amend:  2
age:  35

1. name: John Smith
2. age: 35

OK or number to amend:  OK
val T = (IOSTREAMS(-, -), ["John Smith", "35"]) :
iostreams * string list
```

Of course, in the final version, all the component functions would be placed in a local expression.

Discussion

The new version of *get_tuple* is much more complex than the old one, though the function it performs is the same — it gets a tuple. There is a trade-off here between the amount of time and energy we are prepared to spend refining *get_tuple* and the user-friendliness of the resulting system. For an initial prototype, the first version of *get_tuple*, which is easily manufactured from standard components, would probably suffice. Later on, the more sophisticated 'hand-crafted' version could be appropriate. For a production system, we might decide to invest in more elaborate hardware and produce a version that used cursor control to move from item to item on the screen and update values in place, rather than repeatedly output the current values. The important point is that ALL these possible versions of *get_tuple* will plug into the system via the same interface, that given by the type of the function:

get_tuple : ((*string list* → *string*) * *string*) *list* → *iostreams* → *iostreams* * *string list*

and so each of them can be plugged into the system without any other changes being necessary.

11.4 LIBRARY SYSTEM PROTOTYPE

Where shall we start in designing our library prototype? We could write a function to perform a particular action (say, lending a book) and continue to add more until the prototype is finished, but unfortunately it is very hard to imagine the *exact* form of any particular function until the whole prototype, with its myriad interactions between functions, has been completed. We are in a chicken and egg situation; to untie the knot we have to make an imaginative leap and complete the prototype in our mind before a word of code has been written.

Of course, completing the prototype in our mind may involve writing an exploratory function to see how our ideas work out in practice, and once we start writing functions we will no doubt modify our original ideas, but we do require a goal, however hazy, towards which to work.

First Thoughts

If we imagine a library with a computer terminal installed on the issue desk, we can quickly think of functions that it would be useful for: issuing loans and noting returned books, calculating fines for overdue books, telling would-be borrowers that they have reached their limit, and other unfriendly bureaucratic functions. On the brighter side, a computer terminal installed in the library itself could search for books by author, title or keyword, and tell the inquirer whether the book sought was in the library or on loan. If a required book was on loan, the computer could reserve it for the prospective borrower when it is returned.

The library managers could have a terminal, which they could use to find out which books were frequently borrowed, so that extra copies could be ordered, and infrequently borrowed books could be relegated to a special annexe to make more room on the main shelves. The possibilities of our system are already quite extensive, and no doubt more ideas will be dreamt up once the system is in use. But we still have to write the system!

I suggest that we start with the minimum system that makes any sense, which is one where books can be borrowed and returned. This system will have the following functions:

- Adding new books to the library
- Adding new users to the library
- Lending books to users and allowing users to return books.

Data Structures

Now we have some idea of the functions we are performing, let's think about the data structures we need. We will require a means of storing information about books, and the relational database we invented earlier in this chapter can be used for this: typical attributes for a book will be its title, author and ISBN; we could also hold information on the number of copies held. For users, we will require some personal details such as name, and some unique number (library number) so that a user can be unambiguously identified.

Data about loans will have to be stored somewhere too. Here our relational ideas can help us. A loan can be considered to be a (temporary) relation between a user and a book, so we can model a user borrowing a book by adding the appropriate tuple to our relation *loan*. Similarly, the act of returning a book can be modelled by removing the appropriate tuple from the relation. What should the tuple contain? All we need is some way of unambiguously identifying the relevant user and book; these two numbers will suffice for our loan information. We have a unique number for the user; we must ensure that each book has a unique number too. This can be done by appending a copy number to the ISBN number.

As the reader may have noticed, there is a certain ambiguity about my use of the word 'book'. Do I mean the abstract book, which has a title, an author, and so on, or do I mean one particular copy of the book, which can be borrowed by a user and read? To clarify things, I will refer to the abstract concept as a book, and the actual physical volume as a copy. A book will be unambiguously identified by its ISBN number, while a copy needs an additional copy number to identify it. This means we now need two relations in our database (if we want to avoid massive redundancy of data): the relation *book* will hold details of books in the abstract and will have one tuple for each abstract book, while the relation *copy* will hold details for copies of books, and will have one tuple for each copy. Typically, a copy tuple will just hold the ISBN and the copy number.

Now we are in a position to design our database. The relation *book* can have the following layout

ISBN	*title*	*author*	*publisher*
0-123-45678-9	*Fundamentals of ML*	Cloggs	Triggs & Trogs
9-876-54321-0	*A Short Guide to ML*	Bloggs	Klopf Verlag

while the relation *copy* is just

ISBN	*copy_number*
0-123-45678-9	1
0-123-45678-9	2
9-876-54321-0	1

The relation *user* can just contain the user's number and name:

library_number	*surname*	*forename*
BNP0070682	Bosworth	Richard

while the *loan* relation just holds the number of the user and the unique number of the copy of the book on loan:

library_number	*ISBN*	*copy_number*
BNP0070682	9-876-54321-0	1

(Of course, in real life there should be some sort of time-limit on the loan, requiring the date of borrowing to be recorded, but we omit this in the first prototype.)

Representing the Data Structures in ML

Using our newly written relational database, these structures will be represented as string list lists in ML. As we have seen, we often need to specify the position of an attribute in a relation. Using a number for this purpose (as we did when using *filter_tuple*, for example) is fraught with danger, since the same number could represent a different attribute, depending on which relation we are talking about. When we come to modify the program, there is a not insignificant chance that we may become confused. We can obtain a one-to-one relationship by giving each attribute a unique name:

```
val book_isbn = 1
val book_title = 2
val book_author = 3
val book_publisher = 4

val copy_isbn = 1
val copy_copy_number = 2

val user_library_number = 1
val user_surname = 2
val user_forename = 3

val loan_library_number = 1
val loan_isbn = 2
val loan_copy_number = 3
```

add_book

Let us begin with a function to add a new book to the library, as a library without books is as unhappy a place as a pub with no beer. The function will use *add_tuple* to update the *book* and *copy* relations. If the librarian decides to add *n* copies of the book, then one tuple will be added to the *book* relation, and *n* tuples to the *copy* relation. The copies will have copy numbers 1 to *n*. The relevant values will be obtained from the librarian by using *get_tuple*. The function will take the old *book* relation, the old *copy* relation and the streams and produce a new *book* relation, a new *copy* relation and the updated streams, so its type will be

*relation → relation → iostreams → iostreams * relation * relation*

We first write a function to add *n* copies, as this is the only tricky part of the code. It's very nearly a job for *iter*, but the fact that the copy number changes with each copy added makes explicit recursion necessary. The function will take an ISBN number, the number of copies and the old *copy* relation, and will produce a new *copy* relation. The base case occurs when the number of copies is zero. The recurrence relation is that adding *k* copies consists of adding the kth copy to the result of adding *k*-1 copies:

```
fun add_copies isbn 0 copy = copy
|   add_copies isbn k copy =
  add_tuple [isbn,int_string k]
            (add_copies isbn (k-1) copy)
;

val add_copies = fn : string -> int -> relation ->
relation

val C = add_copies "0521390222" 5 [];

val C = [["0521390222", "1"], ["0521390222", "2"],
["0521390222", "3"], ["0521390222", "4"],
["0521390222", "5"]] : relation
```

Writing the *add_book* function is now just a matter of care as all the trickiness is encapsulated in *add_copies*: We have to first obtain the book data from the iostreams using *get_tuple*, then make up the relevant tuples from this data (using analytical and synthetic *let* expressions) and finally update the relations *book* and *copy*. It's good practice too to give the user a message saying the update has been done successfully. We return the updated iostreams and the relations.

```
fun add_book book copy v =
  let
    val (v1,book_data) =
      get_tuple [(valid_isbn_st,"ISBN"),
                 (valid_cl_st,"title"),
                 (valid_cl_st,"author"),
                 (valid_cl_st,"publisher"),
                 (valid_copies,"no. of copies")]
                v
    val [isbn,title,author,publisher,num_copies]
      = book_data
    val book_tuple =
      [isbn,title,author,publisher]
    val book1 = add_tuple book_tuple book
    val copy1 = add_copies isbn num_copies copy
    val v2 = put_string
                "\nBook added to library\n" v1
  in
    (v2,book1,copy1)
  end
;

ML TYPE ERROR - Unbound variable
INVOLVING:  valid_copies
```

It seems we have forgotten the validation functions!

Validation Functions for *add_book*

valid_cl_st is a standard validation function to be found in Appendix 4, but *valid_isbn_st* and *valid_copies* are particular to this application, and have to be specially written. *valid_copies* is the easier to specify; we require only that the number of copies is at least one:

```
fun valid_copies S =
  let
    val copies = cl_int S
      handle Cl_int =>
      raise Verr "not an integer"
  in
    if copies < 1 then raise Verr "less than one"
    else copies
  end
;
```

```
val valid_copies = fn : string list -> int

valid_copies ["x"];

ML EXCEPTION: Verr("not an integer")

valid_copies ["0"];

ML EXCEPTION: Verr("less than one")

valid_copies ["1"];

val it = 1 : int
```

For *valid_isbn_st* we have a complex requirement: each ISBN is a ten-digit number, the last 'digit' of which can be the letter X. To check that the number is correct and no transcription errors have been made, a modulo 11 checksum is calculated on the 10 digits.

However, for our prototype, we shall avoid such complications and just check that we have a ten-digit number:

```
fun valid_isbn_st S =
  let
    val isbn = cl_int S
      handle Cl_int =>
      raise Verr "not an integer"
  in
    if length S <> 10
    then raise Verr "not 10 digits"
    else implode S
  end
;

val valid_isbn_st = fn : string list -> string

valid_isbn_st (explode "123456789a");

ML EXCEPTION: Verr("not an integer")

valid_isbn_st (explode "123456789");

ML EXCEPTION: Verr("not 10 digits")

valid_isbn_st (explode "1234567890");

val it = "1234567890" : string
```

We can always enhance this function as required, secure in the knowledge that it will still interface perfectly with the rest of the system.

New Version of *add_book*

When we resubmit *add_book* to ML we get a surprise:

```
ML TYPE ERROR — Type unification failure
WANTED    :   (string list -> string) * string
FOUND     :   (string list -> int) * string
```

On searching the function diligently, we find that *valid_copies* is the culprit — it produces an integer instead of a string, and *get_tuple* is required to produce a string list. We can easily turn it into *valid_copies_st*:

```
val valid_copies_st =
  string_it valid_copies;

val valid_copies_st = fn : string list -> string
```

but this will give the wrong type for *num_copies* later on, when it is used as an argument for *add_copies*, which needs an integer. To deal with this requirement (without modifying *add_copies*, which has been proved and tested), we can convert the value using *string_int*, a modification of the *cl_int* function of Chapter 10:

```
exception String_int
fun string_int s =
  cl_int(explode s)
  handle Cl_int => raise String_int
;

exception String_int
val string_int = fn : string -> int
```

Our new version of *add_book* now becomes:

```
fun add_book book copy v =
  let
    val (v1,book_data) =
      get_tuple
        [(valid_isbn_st,"ISBN"),
         (valid_cl_st,"title"),
         (valid_cl_st,"author"),
```

```
          (valid_cl_st,"publisher"),
          (valid_copies_st,"no. of copies")]
          v
    val [isbn,title,author,publisher,num_copies_st]
      = book_data
    val book_tuple =
      [isbn,title,author,publisher]
    val num_copies = string_int num_copies_st
    val book1 = add_tuple book_tuple book
    val copy1 = add_copies isbn num_copies copy
    val v2 = put_string
                "\nBook added to library\n" v1
  in
    (v2,book1,copy1)
  end
;

val add_book = fn : relation -> relation ->
iostreams -> iostreams * relation * relation
```

To test the function, we need only consider the normal case, since all the exceptional ones have been tested already:

```
add_book [] [] io;

1. ISBN:  0123456789
2. title:  Fundamentals of ML
3. author:  Bloggs
4. publisher:  Triggs & Trogs
5. no. of copies:  2

OK or number to amend:  OK

Book added to library
val it = (IOSTREAMS(-, -), [["0123456789",
"Fundamentals of ML", "Bloggs", "Triggs &
Trogs"]], [["0123456789", "1"], ["0123456789",
"2"]]) : iostreams * relation * relation
```

add_loan

The function to add a user is left as an exercise; now I want to show how lending a book can be modelled in the system, assuming we have a population of books and users. Of course, in a real library system the librarian would be equipped with a light-pen to read the library number of the user and the copy number of the book. We shall ignore such complications for our prototype, and assume that all data entry is via the terminal.

Lending a book, then, reduces to obtaining a library number and the unique copy identifier (ISBN plus copy number), and adding this tuple to the *loan* relation. Once again we have the problem of validation, in particular the library number. For the prototype we can avoid this problem by allowing any string as a library number. We can re-use our *valid_copies_st* function to validate the copy number.

Now the *add_loan* function is straightforward. It will take the old *loan* relation and the iostreams and return updated versions of each:

```
fun add_loan loan v =
  let
    val (v1,loan_tuple) =
      get_tuple
        [(valid_cl_st,"library number"),
         (valid_isbn_st,"ISBN"),
         (valid_copies_st,"copy number")]
        v
    val loan1 = add_tuple loan_tuple loan
  in
    (v1,loan1)
  end
;
val add_loan = fn : relation -> iostreams ->
iostreams * relation
```

This looks like a quite elegant function, but even before we test it we can see that it has deficiencies. There is no check that the copy being loaned actually exists in the library. There is no check that the library number belongs to a valid user of the library. Finally, there is no check that the copy has not been loaned out already.

The reader may object that some of these checks are redundant. If the copy being loaned does not exist in the library, how is it that the user is presenting it to the librarian to be loaned? If the user has a library card with a machine-readable number on it, surely that user is a member of the library? And why would a user attempt to borrow a book twice? Does it matter if the attempt is made?

The answers to these questions bring out an interesting point about computerized systems such as this: the need to make a distinction between the system as modelled inside the computer and the actual system in the real world outside the computer. These two systems should mirror one another exactly; if they don't, then something is wrong somewhere. Our checks are a way of giving an early warning that the one-to-one mapping between the internal and external systems is breaking down, and that something needs to be done to rectify the situation. This may involve changes to the internal system, the external system or both; the point is that something must be done.

So these issues must be addressed, but where can we address them? As they are validation issues it seems sensible to address them inside the validation functions. This tactic has the additional advantage of preserving the form of *add_loan* as we have written it.

Loan Validation Revisited

Let us begin with validating the user number. We now have the requirement that the user should have a number that is already known to the system. This condition can be checked by searching the *user* relation for a tuple containing this number, using our *find_tuple* function. Failure to find such a tuple (the exception condition) implies the failure of the validation. The validation function can take the *user* relation as a curried argument.

```
fun valid_user_number_st user S =
  let
    val [un,us,uf] =
      find_tuple user_library_number
                 (implode S)
                 user
      handle Find_tuple =>
        raise Verr "not a known user"
  in
    un
  end
;

val valid_user_number_st = fn : relation -> string
list -> string

valid_user_number_st
  [["1234567890","Fred","Bloggs"]]
  (explode "1234567890")
;

val it = "1234567890" : string
```

```
valid_user_number_st
  [["1234567890","Fred","Bloggs"]]
  (explode "1234567891")
;

ML EXCEPTION: Verr("not a known user")
```

Now for the ISBN and copy number. We can check for a known ISBN in a way analogous to the function we have just written, using the *book* relation instead of the *user* relation. Once we have validated the ISBN, we can use that as an argument in the function to check the copy number against the *copy* relation:

```
fun valid_known_isbn_st book S =
  let
    val [bi,bt,ba,bp] =
      find_tuple book_isbn
                 (implode S)
                 book
      handle Find_tuple =>
        raise Verr "not a known book"
  in
    bi
  end
;

val valid_known_isbn_st = fn : relation -> string
list -> string

valid_known_isbn_st
  [["0-123-45678-9", "Fundamentals of ML",
    "Cloggs", "Triggs&Trogs"]]
  (explode "0-123-45678-9")
;

val it = "0-123-45678-9" : string

valid_known_isbn_st
  [["0-123-45678-9", "Fundamentals of ML",
    "Cloggs", "Triggs&Trogs"]]
  (explode "0-123-45678-8")
;

ML EXCEPTION: Verr("not a known book")
```

(Note that this check includes by implication all the other validity checks for ISBNs, as only valid ISBNs are held in the *book* relation.)

The *valid_known_copy_st* function is given the ISBN, the *copy* relation and the putative copy number. From this information it can construct a copy tuple and check whether it is in the *copy* relation:

```
fun valid_known_copy_st isbn copy S =
  let
    val cn = implode S
  in
    if is_member_relation copy [isbn,cn]
    then cn
    else raise Verr "not a known copy"
  end
;

val valid_known_copy_st = fn : string -> relation
-> string list -> string

valid_known_copy_st
  "0-123-45678-9"
  [["0-123-45678-9","1"]]
  ["1"]
;

val it = "1" : string

valid_known_copy_st
  "0-123-45678-9"
  [["0-123-45678-9","1"]]
  ["2"]
;

ML EXCEPTION: Verr("not a known copy")
```

Now we turn to the last validation issue — whether the book has already been loaned. This is again a function of the copy identifier (ISBN and copy number) along with the *loan* relation. We can write it as a validation function with the ISBN and the *loan* relation as curried arguments:

```
fun valid_unloaned_copy_st isbn loan S =
  let
    val cn = implode S
    val this_isbn =
      filter_tuple loan_isbn isbn loan
    val this_copy =
      filter_tuple loan_copy_number cn this_isbn
  in
    if is_null this_copy then cn
    else raise Verr "already on loan"
  end
;

val valid_unloaned_copy_st = fn : string ->
relation -> string list -> string

valid_unloaned_copy_st
  "0-123-45678-9"
  [["BNP0070682","0-123-45678-9","1"]]
  ["2"]
;

val it = "2" : string

valid_unloaned_copy_st
  "0-123-45678-9"
  [["BNP0070682","0-123-45678-9","1"]]
  ["1"]
;

ML EXCEPTION: Verr("already on loan")
```

Now the total check on the copy number is that it represents a book both known and unloaned:

```
fun valid_loanable_copy_st isbn copy loan S =
  let
    val cn = valid_known_copy_st isbn copy S
    val cn = valid_unloaned_copy_st isbn loan S
  in
    cn
  end
;

val valid_loanable_copy_st = fn : string ->
relation -> relation -> string list -> string
```

Revised Version of *add_loan*

We are at last in a position to write an updated version of *add_loan*, using the revised validation functions:

- *valid_user_number_st user S*
- *valid_known_isbn_st book S*
- *valid_loanable_copy_st isbn copy loan S*

As these validation functions refer to the four relations: *book*, *copy*, *user* and *loan*, the new *add_loan* function must take all of these as arguments. Its interface to the rest of the system has been changed, simply because it does a better job of checking the user's data. We have to accept that our new *add_loan* is a different creature from the old one; for a given state of the system (as represented by the four relations *book*, *copy*, *user* and *loan*) it will behave differently from the old version (and, we hope, in a more acceptable way). We see that there is nothing sacrosanct about the type of a function, and that our first conjecture as to a function's type may be wildly inaccurate.

```
fun add_loan book copy user loan v =
  let
    val (v1,loan_tuple) =
      get_tuple
  [(valid_user_number_st user,"library number"),
   (valid_known_isbn_st book,"ISBN"),
   (valid_loanable_copy_st isbn copy loan,
      "copy number")
  ]
  v
    val loan1 = add_tuple loan_tuple loan
  in
    (v1,loan1)
  end
;

ML TYPE ERROR - Unbound variable
INVOLVING:  isbn
```

Still our labours with this function are not ended! The *valid_loanable_copy_st* function requires the ISBN number from the previous prompt, but, as written, *get_tuple* cannot supply it. What are we to do? We do not want to abandon *get_tuple* with its pleasant user interface, but we do want to validate the copy number properly. One way out of our difficulties would be to do a partial validation of the copy number inside *get_tuple* (just using *valid_copies_st*, which requires no ISBN), and then check the copy identifier when both ISBN and copy number have been successfully captured. If this check fails (indicating an error in the ISBN or the copy number) all the data must be re-entered:

```
fun add_loan book copy user loan v =
  let
    val (v1,loan_tuple) =
      get_tuple
   [(valid_user_number_st user,"library number"),
    (valid_known_isbn_st book,"ISBN"),
    (valid_copies_st, "copy number")
   ]
      v
    val [lu,li,lc] = loan_tuple
    val lc1 = valid_loanable_copy_st li copy
                  loan (explode lc)
      handle Verr s =>
        let
          val v2 =
            put_string
              ("\nCopy is " ^ s ^ "\n\n")
              v1
        in
            add_loan book copy user loan v2
        end
    val loan1 = add_tuple loan_tuple loan
    val v2 =
      put_string "\nCopy successfully loaned\n\n"
                  v1
  in
    (v2,loan1)
  end
;
```

```
ML TYPE ERROR - Type unification failure
WANTED   :  relation -> relation -> relation ->
relation -> iostreams -> iostreams * relation
FOUND    :  relation -> relation -> relation ->
relation -> iostreams -> string
```

ML objects to our little ploy, and for a subtle reason. The handle clause produces a result

add_loan book copy user loan v2

which, by the semantics of handle, must have the same type as

valid_loanable_copy_st li copy loan (explode lc)

i.e. *string*. So ML has inferred two incompatible types for the result of *add_loan*. Something is wrong somewhere, but what is it? We have correctly worked out that an exception from *valid_loanable_copy* requires the *add_loan* function to be applied to the new iostreams (*v2*), but we then attempted to associate the result of this application with the result of *valid_loanable_copy*. The handle clause is in the wrong place — we need to associate the result of the handle clause with the result of applying the original *add_loan*. The combined expression will then have the correct type (*iostreams * relation*):

```
fun add_loan book copy user loan v =
  let
    val (v1,loan_tuple) =
      get_tuple
   [(valid_user_number_st user,"library number"),
    (valid_known_isbn_st book,"ISBN"),
    (valid_copies_st, "copy number")
   ]
      v
    val [lu,li,lc] = loan_tuple
    val lc1 = valid_loanable_copy_st li copy
                loan (explode lc)
    val loan1 = add_tuple loan_tuple loan
    val v2 =
      put_string "\nCopy successfully loaned\n\n"
                 v1
  in
    (v2,loan1)
  end
  handle Verr s =>
  let
    val v2 =
      put_string
        ("\nCopy is " ^ s ^ "\n\n")
        v1
  in
    add_loan book copy user loan v2
  end
```

```
;

ML TYPE ERROR - Unbound variable
INVOLVING:  v1
```

Now we have another problem — the value of the iostreams at the point where the exception occurs (*v1*) has gone out of scope by the time we handle the exception! We are in a hole, and the first rule in this situation is to stop digging. Let us sit back, review the situation, and see if we can improve the code, at least to the point where ML accepts it.

We will return to the original specification of the function and attempt to redesign it, proving as we go. We require a function *add_loan* which will capture the loan information and add a new tuple to the *loan* relation. To validate the loan data, it must check against the *book* relation and the *copy* relation. If the loan information is found to be incorrect, which cannot be known until all the data has been captured, it must ask the user all over again for the data. We have here a recursive function where the base case corresponds to correct entry of the data and the inductive case involves a series of erroneous attempts at entering the data, followed by a good attempt. We require a function that will distinguish these cases, say *is_loanable_copy*, leading to a structure such as:

```
fun add_loan book copy user loan v =
  let
    val (v1,loan_tuple) =
      get_tuple
   [(valid_user_number_st user,"library number"),
    (valid_known_isbn_st book,"ISBN"),
    (valid_copies_st, "copy number")
   ]
      v
    val [lu,li,lc] = loan_tuple
  in
    if is_loanable_copy li copy loan (explode lc)
    then
      let
        val loan1 = add_tuple loan_tuple loan
        val v2 =
          put_string
            "\nCopy successfully loaned\n\n"
            v1
      in
        (v2,loan1)
      end
```

```
      else
        let
          val v2 =
            put_string
              ("\nCopy not loaned\n\n")
            v1
        in
          add_loan book copy user loan v2
        end
    end
;
```

In fact we could do a little better than this, and write a function *loanable_copy_status* which will return a string indicating exactly why the book isn't loanable (or 'OK' if it is). The function will take the *loan* and *copy* relations, the ISBN and the copy number, and return a string:

```
fun loanable_copy_status
  copy loan isbn copy_num =
  let
    val xcn = explode(copy_num)
    val cn = valid_known_copy_st isbn copy xcn
    val cn = valid_unloaned_copy_st isbn loan xcn
  in
    "OK"
  end
  handle Verr s => s
;

val loanable_copy_status = fn : relation ->
relation -> string -> string -> string
```

Now at last we can write our fifth attempt at *add_loan*:

```
fun add_loan book copy user loan v =
  let
    val (v1,loan_tuple) =
      get_tuple
   [(valid_user_number_st user,"library number"),
    (valid_known_isbn_st book,"ISBN"),
    (valid_copies_st, "copy number")
   ]
      v
    val [lu,li,lc] = loan_tuple
```

```
    val status = loanable_copy_status
                     copy loan li lc
  in
    if status = "OK" then
      let
        val loan1 = add_tuple loan_tuple loan
        val v2 =
          put_string
            "\nCopy successfully loaned\n\n"
            v1
      in
        (v2,loan1)
      end
    else
      let
        val v2 =
          put_string
            ("\nCopy is " ^ status ^ "\n\n")
            v1
      in
        add_loan book copy user loan v2
      end
  end
;

val add_loan = fn : relation -> relation ->
relation -> relation -> iostreams -> iostreams *
relation
```

Well, we've proved it, ML has accepted it, now let's test it (with bated breath):

```
add_loan
  [["0-123-45678-9", "Fundamentals of ML",
    "Cloggs", "Triggs&Trogs"]]
  [["0-123-45678-9","1"]]
  [["1234567890","Fred","Bloggs"]]
  []
  io
;

 1. library number:  1234567890
 2. ISBN:  0-123-45678-9
 3. copy number:  1
```

```
 OK or number to amend:  OK

 Copy successfully loaned

val it = (IOSTREAMS(-, -), [["1234567890", "0-123-
45678-9", "1"]]) : iostreams * relation


add_loan
  [["0-123-45678-9", "Fundamentals of ML",
    "Cloggs", "Triggs&Trogs"]]
  [["0-123-45678-9","1"]]
  [["1234567890","Fred","Bloggs"]]
  [["1234567890", "0-123-45678-9", "1"]]
  io
;

 1. library number:  1234567890
 2. ISBN:  0-123-45678-9
 3. copy number:  1

 OK or number to amend:  OK

 Copy is already on loan

 1. library number:  1234567890
 2. ISBN:  0-123-45678-9
 3. copy number:  2

 OK or number to amend:  OK

 Copy is not a known copy

 1. library number:
```

We have arrived at an interesting situation in testing this function. There is now no input which will cause the function to terminate. There is only one book in the library, and this has been borrowed already. *Any* attempt to borrow will be invalid. Fortunately, ML allows for such situations with the *Interrupt* exception, which is raised by 'external intervention', usually by pressing CNTRL and C simultaneously on the keyboard:

```
ML EXCEPTION: Interrupt
```

This brings our testing to a merciful close.

Once again we see the trade-off between brevity and relevance in our prototype; it is very easy to build an initial version of *add_loan* which goes through the motions of lending a book, but doesn't tackle the thorny problem of erroneous input, with the result that the integrity of the database could be seriously compromised. On the other hand, if we do try to take account of some of the difficulties which may arise in practice, we end up with a much more complex prototype. In any given situation, we must use our judgement in deciding just how far to go in our quest for realism.

Persistence of Library Data

The function for returning books is left as an exercise — what I want to do now is show how the entire library system can be put together into one function, while retaining the independence of its parts, and allowing these parts to be modified as required.

If we think of the library system as a function, the first question that arises is its type — what does it take as an argument and return as a result? Well, what is going on when we execute such a function? We interact as a user, via the iostreams, with the values stored in the library, and we modify those values. At the end of the function's operation, the library data will have been modified, and so will the iostreams. So the library function's type will be something like

*library_data → iostreams → iostreams * library_data*

However, there is an additional requirement — when we run the library system again tomorrow, we want the data to be the same as when we finished running it today. The persistence of the library data is longer than any one execution of the function. We need a structure for the library data that takes this requirement into account.

The structure chosen is the stack. The relational database system provides a set of named stacks, implemented on the secondary (persistent) storage of the computer. Functions are provided to push a relation onto a named stack, and to copy a relation from the top of a named stack. No functions are provided to access the other items in the stack, corresponding to previous incarnations of the relation, so these previous incarnations are lost (in the normal operation of the system — they may be needed if a hardware or software error occurs).

In this way the library function, when executed, can always operate on the most up-to-date version of the library data. It simply copies each relation it needs from the appropriate stack, at the start of its execution, and pushes the updated version back to the same stack when it finishes execution. The stacks are deemed to have an initial value consisting of one empty relation, so the problem of initializing the library system is largely solved. (If it is required to re-initialize the data, a subfunction can be provided inside the library system for that purpose.)

The two functions have the following specification:

copy_relation : *string* → *relation*

copy_relation s = *SS*

copy_relation copies the top relation from the stack named by *s* and returns it as *SS*.

push_relation : *relation* → *string* → *string*

push_relation SS s = *s*

push_relation pushes the relation *SS* onto the stack named by string *s*, and returns the name of the stack.

Full details of the functions, and their implementation in the ML input/output system, are given in Appendix 6.

The *Library* Function

We are now in a position to put the library system together as one function, which we can call *library*. Its type is

library_data → *iostreams* → *iostreams* * *library_data*

where *library_data* is a tuple of stack names on secondary storage. Its operation will consist of copying the latest version of the library data from the stacks, giving the user a choice of operations to modify the data, and finally pushing the data back to the stacks when the user decides to call it a day. The selection of operations is a natural task for our *select* function of the previous chapter. If you recall, this function had the following specification:

select : *string* → *string list* → (*iostreams* * *'a* → *iostreams* * *'a*) *list* → *iostreams* * *'a* → *iostreams* * *'a*

It takes a title string, a list of options, and a list of corresponding functions (of type *iostreams* * *'a* → *iostreams* * *'a*) to give a function of type *iostreams* * *'a* → *iostreams* * *'a*. In the case of the library, all the *'a*s are going to be library data in its internal form, i.e. the four relations *book*, *copy*, *user* and *loan*. The functions which perform the operations will be slightly modified versions of functions such as *add_book*. The modification is necessary to ensure that every function has the same type, namely *iostreams* * (*relation* * *relation* * *relation* * *relation*) → *iostreams* * (*relation* * *relation* * *relation* * *relation*).

We will now write a version of *library* for the functions we mentioned earlier this chapter: adding new books and new users to the library, lending books to users, and allowing users to return books. Some of these functions we have implemented already, some of them, dear reader, are to be implemented by you in the exercises at the end of this chapter. We shall represent these for the moment by dummy functions. We have:

```
fun option_add_book (v,(book,copy,user,loan)) =
  let
    val (v1,book1,copy1) = add_book book copy v
  in
    (v1,(book1,copy1,user,loan))
  end
;

val option_add_book = fn : iostreams * (relation *
relation * 'a * 'b) -> iostreams * (relation *
relation * 'a * 'b)


fun option_add_user (v,(book,copy,user,loan)) =
  (v,(book,copy,user,loan))
;

val option_add_user = fn : 'a * ('b * 'c * 'd *
'e) -> 'a * ('b * 'c * 'd * 'e)


fun option_add_loan (v,(book,copy,user,loan)) =
  let
    val (v1,loan1) =
      add_loan book copy user loan v
  in
    (v1,(book,copy,user,loan1))
  end
;
```

```
val option_add_loan = fn : iostreams * (relation *
relation * relation * relation) -> iostreams *
(relation * relation * relation * relation)
```

```
fun option_delete_loan (v,(book,copy,user,loan)) =
(v,(book,copy,user,loan));
```

```
val option_delete_loan = fn : 'a * ('b * 'c * 'd *
'e) -> 'a * ('b * 'c * 'd * 'e)
```

The main library function can now be written:

```
fun library (book_stack, copy_stack, user_stack,
             loan_stack) v =
  let
    val book = copy_relation book_stack
    val copy = copy_relation copy_stack
    val user = copy_relation user_stack
    val loan = copy_relation loan_stack
    val (v1,(book1,copy1,user1,loan1)) =
      select "Library System"
             ["add a book",
              "add a user",
              "lend a book",
              "return a book"
             ]
             [option_add_book,
              option_add_user,
              option_add_loan,
              option_delete_loan
             ]
             (v,(book,copy,user,loan))
    val book_stack1 =
      push_relation book1 book_stack
    val copy_stack1 =
      push_relation copy1 copy_stack
    val user_stack1 =
      push_relation user1 user_stack
    val loan_stack1 =
      push_relation loan1 loan_stack
```

```
  in
    (v1,(book_stack1,copy_stack1,user_stack1,
         loan_stack1))
  end
;

val library = fn : string * string * string *
string -> iostreams -> iostreams * (string *
string * string * string)


library ("book_stack","copy_stack","user_stack",
         "loan_stack") io;

                Library System

 0. Exit

 1. add a book
 2. add a user
 3. lend a book
 4. return a book

 Please enter your selection: 1

 1. ISBN:  0123456789
 2. title:  Fundamentals of ML
 3. author:  Bloggs
 4. publisher:  Triggs & Trogs

 OK or number to amend:  OK
 no. of copies:  2

 Book added to library


                Library System

 0. Exit

 1. add a book
 2. add a user
 3. lend a book
 4. return a book
```

```
Please enter your selection: 0
val it = (IOSTREAMS(-, -), ("book_stack",
"copy_stack", "user_stack", "loan_stack")) :
iostreams * (string * string * string * string)
```

(This of course is not a complete test. All the options should be exercised.)

Discussion

Well, it has been a long and rather tortuous journey, but we got there in the end! Some of the points that emerged on the way:

- When building a prototype, we can't have a detailed picture of the final product. But we need some kind of picture to provide a framework for the functions that we write. Hence the informal design document at the start of this section.
- The design process oscillates between thinking about the functions being performed and the data the functions are acting on. The type system of ML ensures that we specify precisely the type of data each function takes as argument and returns as a result.
- It's easy to build a 'Mickey Mouse' prototype with minimal functionality. Once we attempt to make the prototype more realistic, the complexity of the system increases enormously, in terms of the complexity of both the functions and their interactions. Both *get_tuple* and *add_loan* demonstrated this effect. A compromise has to be found.
- Strictness of typing (especially of constructed types) sometimes has to be sacrificed for practicality of implementation. For example, we chose to make the type of a relation *string list list* rather than have a RELATION constructor, which would have complicated the design of the system and made the logic more obscure, for a small gain in security. ML provides the choice of type binding or type construction to allow us to make the type-checking of our program as strict as we wish.
- A good notation like ML can help us avoid mistakes in design, can sometimes point the way towards a better solution, and can remind us of aspects of the problem that we had forgotten. For example, the strategy of off-loading most of the checking to validation functions, with the consequent increase in clarity of the main functions, was made easier because of the excellent exception-handling capabilities of ML. Again, when we updated *add_loan*, ML reminded us that the ISBN was undefined in the check for the copy number, forcing us to rethink our validation strategy, and also guided us to a more straightforward implementation (by rejecting our initial convoluted ideas!).

- The type-system of ML can sometimes be a little too stringent, as when we had to modify *add_book* merely to obtain type-compatibility within a list. But this is a small price to pay for the benefits of automatic type-checking, as pointed out above.
- The combination of informal proving of the design, type-checking by the interpreter and testing of critical cases gives us a large degree of confidence in our final design.

11.5 SYNTAX INTRODUCED IN THIS CHAPTER

dec	::=	type *typbind*	type declaration
typbind	::=	*tycon* = *ty*	type binding

11.6 CHAPTER SUMMARY

Three case studies have been presented: a simple relational database, a form management system and a prototype system for use in a library. These systems build on the ideas of previous chapters, and the library system is itself built on top of the other two. The simplicity and naturalness of design which was apparent in earlier chapters still manifests itself in these larger-scale examples, but a clear trade-off is observed between realism and elegance. It proved possible, however, to encapsulate most of the difficulties within subfunctions, so that the main functions remained reasonably clear and uncluttered. The system is highly modular and can be easily modified and extended.

The examples of this chapter demonstrate, I hope, that a functional programming language, with a suitable set of support functions, is a strong candidate for the design and implementation of such prototypes.

EXERCISES

1. What are the values of the following expressions?:
   (a) add_tuple [“a”,“b”,“c”] [[“a”,“b”,“c”],[“d”,“e”,“f”]]
   (b) add_tuple [“a”,“b”,“g”] [[“a”,“b”,“c”],[“d”,“e”,“f”]]
   (c) add_tuple [“a”,“b”] [[“a”,“b”,“c”],[“d”,“e”,“f”]]
   (d) delete_tuple [“a”,“b”,“c”] [[“a”,“b”,“c”],[“d”,“e”,“f”]]
   (e) delete_tuple [“a”,“b”,“g”] [[“a”,“b”,“c”],[“d”,“e”,“f”]]
2. Write a function *hi_attribute* which, given an attribute number and a relation, finds the highest value of that attribute in the relation. The function will raise exception *Hi_attribute* on null relation or invalid attribute number.
3. And now the complementary function *lo_attribute*.
4. Write a function *product* which forms the cartesian product of two relations. The cartesian product contains all the possible combinations of tuples in the two relations, so if we have a relation *titles*:

title	*ISBN*
Fundamentals of ML	0-123-45678-9
A Short Guide to ML	9-876-54321-0

and another relation *authors*:

author
Bloggs
Cloggs

then *product*(*titles*,*authors*) will have the value

title	*ISBN*	*author*
Fundamentals of ML	0-123-45678-9	Bloggs
Fundamentals of ML	0-123-45678-9	Cloggs
A Short Guide to ML	9-876-54321-0	Bloggs
A Short Guide to ML	9-876-54321-0	Cloggs

Use the conventions of this chapter to represent a relation as a *string list list*.

5. Design a natural *join* operator for two relations. For the *join* operation, an attribute is selected which is common to each relation, and a set of tuples is selected from the cartesian product of the relations for which the two attributes have the same value. Then the second attribute column (which is of course identical to the first) is dropped from the resulting relation. So suppose we have a relation *books*:

title	*ISBN*	*author*
Fundamentals of ML	0-123-45678-9	Bloggs
Fundamentals of ML	0-123-45678-9	Cloggs
A Short Guide to ML	9-876-54321-0	Bloggs

and another relation *publishers*:

ISBN	*publisher*
0-123-45678-9	Triggs & Trogs
9-876-54321-0	Klopf Verlag

then *join* (2, *books*, 1, *publishers*) would have the value

title	*ISBN*	*author*	*publisher*
Fundamentals of ML	0-123-45678-9	Bloggs	Triggs & Trogs
Fundamentals of ML	0-123-45678-9	Cloggs	Triggs & Trogs
A Short Guide to ML	9-876-54321-0	Bloggs	Klopf Verlag

6. Modify *valid_isbn_st* so that it performs a modulo 11 checksum on the ISBN. The ISBN consists of nine decimal digits and a checksum character which can be digits 0–9 or X (representing 10).
   The checksum is calculated as follows: multiply the rightmost (i.e. check) digit by 1, the next digit by 2 and so on, ending up multiplying the leftmost digit by 10. Add the results of these multiplications together, and divide by 11. The remainder should be zero.
7. Modify *add_book* so that it checks that the ISBN of the book being added does not already exist in the library.
8. Design and write the function *add_user*, which adds a new user to the library.
9. Design and write the function *delete_loan*, which is activated when a user returns a book.

APPENDIX

ONE

ML STANDARD FUNCTIONS

Here is a list of some standard functions available in ML:

`not : bool -> bool`	inverts a Boolean value, e.g. *not(true) = false*
`~ : num -> num`	negates a number, or exception *N*um if the result is out of range
`abs : num -> num`	returns the absolute value of a number, e.g. *abs*(~3) = 3, or the exception *Abs* if the result is out of range
`floor : real -> int`	returns the greatest integer which is less than the argument, e.g. *floor* 2.34 = 2, *floor* ~2.34 = ~3, or the exception *Floor* if the result is out of range
`real : int -> real`	returns the real number corresponding to the argument, e.g. *real* 2 = 2.0
`sqrt : real -> real`	returns the square root of the argument, or the exception *Sqrt* if the argument is negative
`sin : real -> real`	returns the sine of the argument considered as radians
`cos : real -> real`	returns the cosine of the argument considered as radians

`arctan : real -> real` returns the angle in radians (in the range $\pm \pi/2$) whose tangent is the argument (i.e. $\tan^{-1}$)

`exp : real -> real` returns e^x where x is the argument, or exception *Exp* if the result is out of range

`ln : real -> real` returns the natural logarithm of the argument, or the exception *Ln* if the result is out of range

`size : string -> int` returns the length in characters of a string

`chr : int -> string` returns the ith character in the ASCII character set as a 1-character string, where i is the argument, or exception *Chr* if i is not in the interval [0, 255]

`ord : string -> int` returns the number of the first character of the argument in the ASCII character set, or exception *Ord* if the argument is the empty string

`explode : string -> string list`
returns the list of characters (as single-character strings) of which the argument consists

`implode : string list -> string`
returns the string formed by concatenating all members of the list of strings given as argument

ML STANDARD OPERATORS

Here is a list of some standard infixed operators available in ML:

Precedence 7

`/ : real * real -> real` real division; returns exception *Quot* if the result is undefined or out of range

`div : int * int -> int` integer division, e.g. *7 div 3 = 2*; returns exception *Div* if the second argument is zero

`mod : int * int -> int`	modulo, e.g. 7 *mod* 3 = 1; returns exception *Mod* if the second argument is zero
`* : num * num -> num`	multiplication; returns exception *Prod* if the result is undefined or out of range

Note that *mod* and *div* are defined such that the remainder always has the same sign as the divisor, e.g. ~3 div 2 = ~2, ~3 mod 2 = 1.

Precedence 6

`+ : num * num -> num`	addition; returns exception *Sum* if the result is undefined or out of range
`- : num * num -> num`	subtraction; returns exception *Diff* if the result is undefined or out of range
`^ : string * string -> string`	string concatenation

Precedence 5

`:: : 'a * 'a list -> 'a list`	prefixes a value to a list
`@ : 'a list * 'a list -> 'a list`	concatenates two lists

Precedence 4

`= : ''a * ''a -> bool`	compares two values of an equality type for equality
`<> : ''a * ''a -> bool`	compares two values of an equality type for inequality
`< : num * num -> bool`	less than
`> : num * num -> bool`	greater than
`<= : num * num -> bool`	less than or equal to
`>= : num * num -> bool`	greater than or equal to

APPENDIX
TWO
THE ASCII TABLE OF CHARACTERS

To produce a table of the printable ASCII characters with their associated codes, we require a model in ML of the table. Each entry in the table will be of the form (code, associated character), or in ML-ese (*i*, *chr i*), and we can model the entire table as a list of these pairs. Its type will be (*int * string*) *list*.

Let's imagine we are producing an arbitrary subsection of this table (i.e. an arbitrary sublist) running from code *m* to code *n*, inclusive. We can design a function *ascii_tab* to produce this sublist, using the method of differences. The type of the function will be $int \rightarrow int \rightarrow (int * string)\ list$.

For the base case, a null list, we have the condition that $m > n$. For the inductive case, when $m \leq n$, we can imagine prefixing a new pair to the front of the sublist. This pair will have a code one less than the first pair in the sublist, giving the recurrence relation

$$ascii\_tab\ m\ n \ = \ (m, chr\ m) :: ascii\_tab\ (m{+}1)\ n$$

We put the two cases together to obtain the ML function:

```
fun ascii_tab m n =
  if m > n then []
  else (m,chr m) :: ascii_tab (m+1) n
;

val ascii_tab = fn : int -> int -> (int * string)
list
```

Assuming that we know by some miraculous means that the printable portion of the ASCII table runs from code 32 to code 126, we can produce the list corresponding to this portion by saying

```
val printable_ascii_tab = ascii_tab 32 126;
val printable_ascii_tab =
[(32, " "), (33, "!"), (34, "\""), (35, "#"),
 (36, "$"), (37, "%"), (38, "&"), (39, "'"),
 (40, "("), (41, ")"), (42, "*"), (43, "+"),
 (44, ","), (45, "-"), (46, "."), (47, "/"),
 (48, "0"), (49, "1"), (50, "2"), (51, "3"),
 (52, "4"), (53, "5"), (54, "6"), (55, "7"),
 (56, "8"), (57, "9"), (58, ":"), (59, ";"),
 (60, "<"), (61, "="), (62, ">"), (63, "?"),
 (64, "@"), (65, "A"), (66, "B"), (67, "C"),
 (68, "D"), (69, "E"), (70, "F"), (71, "G"),
 (72, "H"), (73, "I"), (74, "J"), (75, "K"),
 (76, "L"), (77, "M"), (78, "N"), (79, "O"),
 (80, "P"), (81, "Q"), (82, "R"), (83, "S"),
 (84, "T"), (85, "U"), (86, "V"), (87, "W"),
 (88, "X"), (89, "Y"), (90, "Z"), (91, "["),
 (92, "\\"), (93, "]"), (94, "^"), (95, "_"),
 (96, "`"), (97, "a"), (98, "b"), (99, "c"),
 (100, "d"), (101, "e"), (102, "f"), (103, "g"),
 (104, "h"), (105, "i"), (106, "j"), (107, "k"),
 (108, "l"), (109, "m"), (110, "n"), (111, "o"),
 (112, "p"), (113, "q"), (114, "r"), (115, "s"),
 (116, "t"), (117, "u"), (118, "v"), (119, "w"),
 (120, "x"), (121, "y"), (122, "z"), (123, "{"),
 (124, "|"), (125, "}"), (126, "~")
] : (int * string) list
```

APPENDIX
THREE
CHARACTER LIST FUNCTIONS

```
local

val sig_digs = 6;
(* no of significant digits for real numbers *)


(* unfoldl : ('a -> bool -> 'b * 'a) -> 'a -> 'b
list *)
(*
unfold f a = B
unfolds composite value a from left to right using
function f to give list B.
Typical call of f is  f a = (p,b,a').  f analyses
a into two parts a' and b, where a' is a "smaller"
version of a. If the analysis is not possible, p
is false and a' and b are undefined.
*)
fun unfoldl f a =
  let
    val (p,b,a') = f a
  in
    if p then b :: unfoldl f a'
         else []
  end
;
```

```
(* unfoldr : ('a -> bool -> 'a * 'b) -> 'a -> 'b
list *)
(*
unfoldr f a = B
unfolds composite value a from right to left using
function f to give list B.
Typical call of f is  f a = (p,a',b).  f analyses
a into two parts a' and b, where a' is a "smaller"
version of a. If the analysis is not possible, p
is false and a' and b are undefined.
*)
fun unfoldr f a =
  let
    fun unfoldr' f a B =
      let
        val (p,a',b) = f a
      in
        if p then unfoldr' f a' (b::B)
             else B
      end
  in
    unfoldr' f a []
  end
;



(* val dig_lis_int : string list -> int *)
(*
dig_lis_int S = i
converts a list of digit characters to a positive
integer
leading zeroes are ignored, null list has value 0.
This is a partial function - action on non-digit
lists is undefined.
*)
val dig_lis_int =
  foldl (fn (i,s) => i * 10 + ord s - ord "0") 0
;



(* pos_frac_cl : int -> real -> string list *)
(*
```

```
pos_frac_cl n x = S
converts a positive real fraction x to a character
list S of length n.
This is a partial function - action on negative or
non-fractional numbers is und
efined
*)
fun pos_frac_cl n x =
  let
  (* chop_real_frac : int * real -> bool * string
     * (int * real) *)
  (*
  chop_real_frac (n,x) = (p,s,(n',x'))
   chops first digit character off real fraction x
to n sig_digs
  *)
  fun chop_real_frac (n,x) =
  if n = 0 then (false,"",(0,0.0))
  else
  let
    val x10 = x * 10.0
    val dig1 = floor x10
    val rest = x10 - real dig1
  in
    (true,chr(dig1 + ord "0"),(n-1,rest))
  end
in
  unfoldl chop_real_frac (n,x)
end;


(* rem_zeroes : string list -> string list *)
(*
rem_zeroes S = S1
removes trailing zeroes from a character list
*)
fun rem_zeroes []          = []
|   rem_zeroes ("0"::S) =
  let
    val rest = rem_zeroes S
  in
    if rest = [] then [] else "0":: rest
  end
|   rem_zeroes (s::S)    = s::rem_zeroes S
;
```

```
(* split_real_cl : string list -> string list *
string list *)
(*
split_real_cl S = (S1,S2)
splits character list representing real number
into integral and fractional part
s.
*)
exception Split_real_cl
fun split_real_cl [] = raise Split_real_cl
|   split_real_cl ["."] = raise Split_real_cl
|   split_real_cl ("."::S) = ([],S)
|   split_real_cl (s::S) =
let
  val (icl,fcl) = split_real_cl S
in
  (s::icl,fcl)
end
;


(* scale_down : int -> real -> real *)
(*
scale_down n x = x1
scales down x by 10 ** n
*)
val scale_down = compose (fn x => x /10.0);


in

(* functions taking a character as argument *)
(* - - - - - - - - - - - - - - - - - - - - *)

(* val is_digit: string -> bool *)
(* returns true iff character is digit *)
fun is_digit s = s >= "0" andalso s <= "9";

(* val is_lowercase: string -> bool *)
(* returns true iff character is lowercase *)
fun is_lowercase s = s >= "a" andalso s <= "z";

(* val is_uppercase: string -> bool *)
(* returns true iff character is uppercase *)
fun is_uppercase s = s >= "A" andalso s <= "Z";
```

```
(* val is_letter: string -> bool *)
(* returns true iff character is a letter *)
fun is_letter s =
  is_lowercase s orelse is_uppercase s
;

(* val digit_val: string -> int *)
(* returns integer value of digit character *)
exception Digit_val;
fun digit_val s =
  if is_digit s then ord s - ord "0"
  else raise Digit_val
;

(* val int_digit : int -> string *)
(*
returns character corresponding to integer digit
*)
exception Int_digit
fun int_digit i =
  if i < 0 orelse i > 9 then raise Int_digit
  else chr(ord "0" + i)
;


(* functions taking character list as argument *)
(* - - - - - - - - - - - - - - - - - - - - - - *)

(* predicates *)
(* - - - - - - *)

(* val is_null_cl : string list -> bool *)
(*
returns true if argument is null character list
*)
fun is_null_cl S = S = [];

(* val is_digit_cl : string list -> bool *)
(*
returns true iff all characters in character list
are digits
*)
val is_digit_cl = all is_digit;
```

```
(* val is_letter_cl : string list -> bool *)
(*
returns true iff all characters in character list
are letters
*)
val is_letter_cl = all is_letter;



(* number conversion *)
(* - - - - - - - - - - - *)

(* val pos_int_cl : int -> string list *)
(*
converts positive integer to list of digits.
0 converts to ["0"]
*)
exception Pos_int_cl
fun pos_int_cl 0 = ["0"]
|  pos_int_cl n =
  if n < 0 then raise Pos_int_cl
  else
  let
    (* chop_int : int -> bool * int * string *)
    (*
    chops  last  digit  character  off  non-zero
    integer
    *)
    fun chop_int 0 = (false,0,"")
    |   chop_int n =
    let
      val front_n = n div 10
      val last_dig = n mod 10
    in
      (true,front_n,chr(last_dig + ord "0"))
    end
  in
    unfoldr chop_int n
  end
;


(* val int_cl : int -> string list *)
(*
```

```
converts integer to character list
negative integer has "-" as first char in list
(i.e. normal convention not ML )
as this function is intended for use in output
functions
*)
fun int_cl n  =
  if n < 0 then "-"::pos_int_cl (~n)
  else pos_int_cl n
;



(* val cl_pos_int: string list -> int *)
(*
converts a character list of one or more digits to
a positive integer
leading zeroes are ignored
*)
exception Cl_pos_int
fun cl_pos_int S =
  if S = [] orelse not(is_digit_cl S)
  then raise Cl_pos_int
  else dig_lis_int S
;



(* val cl_int : string list -> int *)
(*
converts a character list of one or more digits
preceded by an optional "-" character to an
integer value
*)
exception Cl_int
fun cl_int [] = raise Cl_int
|   cl_int ("-"::cs) =
  (~(cl_pos_int cs)
  handle Cl_pos_int => raise Cl_int)
|   cl_int cs =
  cl_pos_int cs
  handle Cl_pos_int => raise Cl_int
;
```

```
(* pos_real_cl : real -> string list *)
(*
converts a positive real number to a character
list
*)
exception Pos_real_cl
fun pos_real_cl x =
if x < 0.0 then raise Pos_real_cl
else
  let
    val ip = floor x
    val fp = x - real ip
    val first_fcl::rest_fcl = pos_frac_cl sig_digs
    fp
  in
    pos_int_cl ip @
      ("." :: first_fcl :: rem_zeroes(rest_fcl))
  end
;


(* real_cl : real -> string list *)
(*
converts real number to character list.
negative integer has "-" as first char in list
(i.e. normal convention not ML) as
this function is intended for use in output func-
tions
*)
fun real_cl x =
  if x < 0.0 then "-" :: pos_real_cl (~x)
  else pos_real_cl x
;


(* cl_pos_real : string list -> real *)
(*
converts character list to positive real number.
Gives exception Cl_pos_real if conversion is
impossible
*)
exception Cl_pos_real
fun cl_pos_real S =
  let
```

```
    val (icl,fcl) = split_real_cl S
      handle Split_real_cl => raise Cl_pos_real
    val shortfcl = take sig_digs fcl
    val lenfcl = length shortfcl
  in
    real(cl_int icl) +
      scale_down lenfcl (real(cl_int shortfcl))
    handle Cl_int => raise Cl_pos_real
  end
;


(* cl_real : string list -> real *)
(*
converts a string list to a real number.
Gives   exception   Cl_real   if   conversion   is
impossible
*)
exception Cl_real
fun cl_real [] = raise Cl_real
|   cl_real (" - "::S) =
  (~(cl_pos_real S)
    handle Cl_pos_real => raise Cl_real)
|   cl_real S =
  (cl_pos_real S
    handle Cl_pos_real => raise Cl_real)
;


(* justification *)
(* - - - - - - - - *)

(*  val  right_justify  :  int  ->  string  list  ->
string list *)
(*
right_justify n S = S1
right justifies character list S in character list
S1 of minimum length n,
using spaces to fill
*)
fun right_justify n S =
  list_of " " (n - length S) @ S
;
```

```
(* val left_justify : int -> string list -> string
list *)
(*
left_justify n S = S1
left justifies character list S in character list
S1 of minimum length n, using
spaces to fill
*)
fun left_justify n S =
  S @ list_of " " (n - length S)
;

end (* local declaration *);
```

SYNTAX INTRODUCED IN THIS APPENDIX

Comment, as can be seen, is written between the **comment brackets** (* and *). Comments can be **nested**, that is,

```
(* comment is written between (* and *) *)
```

is valid comment. Comment is completely ignored by the ML system, and is for the benefit of human readers only.

APPENDIX FOUR

VALIDATION FUNCTIONS

Note that this appendix uses functions from Appendix 3.

```
exception Verr of string

local

(* string_it : ((string list -> 'a)
                -> (string list -> string)
string_it vf
converts a common validation function to one which
returns a string
*)
fun string_it f S =
  let
    val result = f S
  in
    implode S
  end
;


in


(* valid_cl : string list -> string list *)
fun valid_cl S = S;
```

```
(* valid_cl_st : string list -> string *)
val valid_cl_st = string_it valid_cl;


(* valid_digits : string list -> string list *)
fun valid_digits S =
  if is_digit_cl S then S
  else raise Verr "not all digits"
;


(* valid_digits_st : string list -> string *)
val valid_digits_st = string_it valid_digits;


(* valid_letters : string list -> string list *)
fun valid_letters S =
  if is_letter_cl S then S
  else raise Verr "not all letters"
;


(* valid_letters_st : string list -> string *)
val valid_letters_st = string_it valid_letters;


(* valid_letter : string list -> string *)
fun valid_letter [s] =
  if is_letter s then s
  else raise Verr "not a letter"
|   valid_letter _   = raise Verr "not a letter"
;


(* valid_char_set : string list -> string list ->
string *)
(*
valid_char_set S T = s
S is non-null list of valid characters
T is validated as an item of S
*)
exception Valid_char_set
fun valid_char_set (S as (s::S1)) [t] =
```

```
  if is_item t S then t
  else raise Verr ("not " ^
    (foldl (fn (s1,s2) => s1 ^ " or " ^ s2) s S1))
|   valid_char_set (s::S1) _   =
  raise Verr "not a valid character"
|   valid_char_set _ _ = raise Valid_char_set
;


(* valid_integer : string list -> int *)
fun valid_integer S =
  cl_int S
  handle Clint =>
    raise Verr "not an integer number"
  ;


(* valid_integer_st : string list -> string *)
val valid_integer_st = string_it valid_integer;


(* valid_range : int -> int -> string list -> int
*)
(*
valid_range m n S = i
checks S is integer in range m..n
*)
fun valid_range m n S =
  let
    val i = cl_int S
    handle Cl_int => raise Verr "not an integer"
  in
    if m <= i andalso i <= n then i
    else raise Verr ("not an integer in range " ^
                     implode(int_cl m) ^ ".." ^
                     implode(int_cl n))
  end
;


(* valid_range_st : int -> int -> string list ->
string *)
fun valid_range_st m n =
  string_it (valid_range m n)
;
```

```
(* valid_real : string list -> real *)
fun valid_real C =
  cl_real C
  handle Cl_real =>
    raise Verr "not a real number"
;


(* valid_real_st : string list -> string list *)
val valid_real_st = string_it valid_real;


(* valid_date : string list -> string list *)
(*
accepts only valid date in format DDMMYY (years
1901-2099 assumed)
*)
fun valid_date S =
  let
    fun isleap n = n mod 4 = 0;

    exception Splitdate
    fun splitdate [c1,c2,c3,c4,c5,c6] =
      ([c1,c2],[c3,c4],[c5,c6])
    |   splitdate cs = raise Splitdate
    ;

    exception Convnum
    fun convnum [c1,c2] =
      10 * digit_val c1 + digit_val c2
    |   convnum cs = raise Convnum
    ;

    fun valid_month n = 1 <= n andalso n <= 12;

    fun numofdays y =
      fn 2 => if isleap y then 29 else 28
       |  4 => 30
       |  6 => 30
       |  9 => 30
       | 11 => 30
       | _  => 31
    ;
```

```
    in
      if length S <> 6
      then raise Verr "not of length 6"
      else if not (is_digit_cl S)
      then raise Verr "not all digits"
      else
        let val (dd,mm,yy) = splitdate S
            val d = convnum dd
            val m = convnum mm
            val y = convnum yy
        in
          if not (valid_month m)
          then raise Verr "has invalid month number"
          else if d < 1 orelse d > numofdays y m
          then raise Verr "has invalid day number"
          else S
        end
    end
;

(* valid_date_st : string list -> (string * vind)
*)
val valid_date_st = string_it valid_date;

end (* local declaration *);
```

APPENDIX

FIVE

PSEUDO-IMPLEMENTATION OF INPUT/OUTPUT FUNCTIONS

```
datatype iostreams =
IOSTREAMS of instream * outstream;


(* val iostreams : instream -> outstream ->
iostreams *)
(*
iostreams is os = v
converts an instream is and outstream os into an
interactive stream pair v
*)
fun iostreams is os = IOSTREAMS(is,os);


(* val get_line : iostreams -> (iostreams * string
list) *)
(*
get_line v = (v1,S)
Removes first text line of input stream and
returns it as a character list S, along with the
updated stream pair
The result is returned as a character list rather
than a string to facilitate analysis by pattern-
matching
```

```
*)
fun get_line (v as IOSTREAMS(is,os)) =
  let
    fun get_line' is =
      let
        val c = input(is,1)
      in
        if c = "\n" then []
        else
          c::get_line' is
  (* _is_ is current value of input stream *)
      end
  in
    (v,get_line' is)
(* v is current value of stream pair *)
  end
;


(* val put_string : string -> iostreams ->
iostreams *)
(*
put_string s v = v1
appends string s to output stream, and returns
updated streams
*)
fun put_string s (v as IOSTREAMS(is,os)) =
  let
    val u = output(os,s)
  in
    v  (* current value of stream pair *)
  end
;

(* by convention: *)

val io = iostreams std_in std_out
```

APPENDIX

SIX

DATABASE FUNCTIONS

The following two functions allow a relation to be copied from an external file stack and pushed onto an external file stack. In this way information can be saved between runs of an ML program.

```
type tuple = string list
type relation = tuple list
type file_stack = string


(* val copy_relation : file_stack -> relation *)
(*
  copy_relation F = R
  copies relation R from top of file stack F
  The file stack remains unchanged.
*)
local
  (* val load_att : instream -> string *)
  (*
  load_att IN = s
  loads attribute s from input stream IN
  *)
  fun load_att IN =
    let val c = input(IN,1)
    in
      if c = "\t" then ""
      else c ^ load_att IN
    end
```

```
  (* val load_tup : instream -> tuple *)
  (*
  load_tup IN = T
  loads tuple T from input stream IN
  *)
  fun load_tup IN =
    if lookahead IN  = "\n" then
      let
        val d = input(IN,1)
      in
        []
      end
    else load_att IN :: load_tup IN

  fun copy_rel IN =
    if end_of_stream IN then []
    else load_tup IN :: copy_rel IN
in
  fun copy_relation (F:file_stack) =
    let
      val IN = open_in F
      val R = copy_rel IN
      val d = close_in IN
    in
      R:relation
    end
    handle Io _ => []
end
;


(* val push_relation : relation -> file_stack ->
file_stack *)
(*
push_relation R F = F1
pushes relation R onto file stack F to give F1
*)
local
  (* val store_tup : outstream -> tuple -> unit *)
  (*
  store_tup OUT T = ()
  stores tuple T on stream OUT
  *)
  fun store_tup OUT [] = output(OUT,"\n")
```

```
  |   store_tup OUT (x::T) =
    (output(OUT,x ^"\t"); store_tup OUT T)

  (* val store_rel : outstream -> relation -> unit
  *)
  (*
  store_rel OUT R = ()
  stores relation R on stream OUT
  *)
  fun store_rel OUT [] = ()
  |   store_rel OUT (T::R) =
    (store_tup OUT T; store_rel OUT R)
in
  fun push_relation (R:relation) (F:file_stack) =
    let
      val OUT = open_out F
      val d = store_rel OUT R
      val dd = close_out OUT
    in
      F:file_stack
    end
end;
```

BIBLIOGRAPHY

Functional Programming

Bird, R. and Wadler, P., *Introduction to Functional Programming*, Prentice Hall, 1988. Uses a lazy functional language similar to Miranda™. Interesting to contrast with the strict style used in the present book, especially in the areas of list processing and input/output.

Harrison, R., *Abstract Data Types in Standard ML*, Wiley, 1993. An intermediate text which gives a thorough treatment of standard ADTs using SML, starting with lists and working though trees and sets to graphs.

Myers, C., Clack, C. and Poon, E., *Programming with Standard ML*, 1993. Covers the Core language and Modules in 300 pages, so recommended for experienced programmers requiring a quick introduction to ML. Gives practical examples and emphasizes good design techniques.

Paulson, L., *ML for the Working Programmer*, Cambridge University Press, 1991. An advanced text covering all of SML, including the non-functional features and ML modules. Aimed at readers who already know an imperative language and some discrete mathematics, it provides a practical guide to writing efficient ML programs.

Ullman, J. D., *Elements of ML Programming*, Prentice Hall, 1994. Aimed at the reader with a working knowledge of an imperative language, this text covers the Core language and Modules, plus additional features found in the New Jersey implementation of SML. A friendly introduction to ML rather than to functional programing as such.

Wikstrom, Å., *Functional Programming Using Standard ML*, Prentice Hall, 1987. The first elementary textbook on ML, it covers the Core language, with a host of examples and exercises.

Logic and Discrete Mathematics

Fenton, N. and Hill G., *Systems Construction and Analysis*, McGraw-Hill, 1993. Presents the mathematics needed by the modern computing practitioner in a user-friendly fashion.

Databases

Korth, H. F. and Silberschatz, A., *Database Systems Concepts*, 2nd edn, McGraw-Hill, 1991. Recommended for those readers who wish to know more about modern database systems.

INDEX

Index of functions